Intersections of
Thick Cantor Sets

Recent Titles in This Series

(See the AMS catalogue for earlier titles)

MEMOIRS
of the
American Mathematical Society

Number 468

Intersections of
Thick Cantor Sets

Roger Kraft

May 1992 • Volume 97 • Number 468 (second of 3 numbers) • ISSN 0065-9266

American Mathematical Society
Providence, Rhode Island

1991 *Mathematics Subject Classification.*
Primary 58F14, 34C37, 54F50, 28A80.

Library of Congress Cataloging-in-Publication Data

Kraft, Roger, 1956–
    Intersections of thick Cantor sets/Roger Kraft.
    p. cm. – (Memoirs of the American Mathematical Society, ISSN 0065-9266; no. 468)
    Includes bibliographical references.
    ISBN 0-8218-2528-3
    1. Cantor sets. 2. Difference sets. 3. Bifurcation theory. I. Title. II. Series.
QA3.A57 no. 468
[QA614.8]
500 s–dc20   92-6942
[515′.352]   CIP

Memoirs of the American Mathematical Society
This journal is devoted entirely to research in pure and applied mathematics.

Subscription information. The 1992 subscription begins with Number 459 and consists of six mailings, each containing one or more numbers. Subscription prices for 1992 are $292 list, $234 institutional member. A late charge of 10% of the subscription price will be imposed on orders received from nonmembers after January 1 of the subscription year. Subscribers outside the United States and India must pay a postage surcharge of $25; subscribers in India must pay a postage surcharge of $43. Expedited delivery to destinations in North America $30; elsewhere $82. Each number may be ordered separately; *please specify number* when ordering an individual number. For prices and titles of recently released numbers, see the New Publications sections of the *Notices of the American Mathematical Society.*

    Back number information. For back issues see the *AMS Catalogue of Publications.*

    Subscriptions and orders should be addressed to the American Mathematical Society, P. O. Box 1571, Annex Station, Providence, RI 02901-1571. *All orders must be accompanied by payment.* Other correspondence should be addressed to Box 6248, Providence, RI 02940-6248.

Memoirs of the American Mathematical Society is published bimonthly (each volume consisting usually of more than one number) by the American Mathematical Society at 201 Charles Street, Providence, RI 02904-2213. Second-class postage paid at Providence, Rhode Island. Postmaster: Send address changes to Memoirs, American Mathematical Society, P. O. Box 6248, Providence, RI 02940-6248.

Table Of Contents

Abstract

Thickness is a positive real number assigned to a Cantor set embedded in the real line. Thickness describes, in some sense, the size of the Cantor set. Two Cantor sets are interleaved if neither Cantor set lies in the closure of a component of the compliment of the other. This paper proves the following theorem. The set of all pairs of positive real numbers (τ, τ') with $\tau\tau' > 1$ can be partitioned into three subsets such that, if Γ and Γ' are interleaved Cantor sets with thicknesses in the first subset, then $\Gamma \cap \Gamma'$ must contain a Cantor set; if Γ and Γ' are interleaved Cantor sets with thicknesses in the second subset, then $\Gamma \cap \Gamma'$ must contain an infinite number of points; and if (τ, τ') is in the third subset, then there exist interleaved Cantor sets Γ and Γ' with thicknesses τ and τ' such that $\Gamma \cap \Gamma'$ contains only one point. We give inequalities that exactly describe these subsets of $\tau\tau' > 1$. These subsets are determined by the dynamics of a dynamical system that is derived from a geometric construction that defines, for any pair of positive numbers (τ, τ') with $\tau\tau' > 1$, two intersecting, compact, perfect subsets of the real line. These intersecting sets are either Cantor sets with thicknesses τ and τ' and with only one point in their intersection, or they are not Cantor sets. The dynamics of the dynamical system determine whether or not these sets are Cantor sets. By analyzing this dynamical system we can describe all possible ways that two interleaved Cantor sets can intersect at only one point.

Key words and phrases. Cantor sets, thickness, difference sets, homoclinic bifurcations.

Introduction

Sheldon Newhouse developed the concept of thickness in order to prove that under certain conditions, two Cantor sets must intersect. Thickness assigns to every Cantor set embedded in the real line a number from $[0, \infty)$. Newhouse showed that if two Cantor sets Γ and Γ' have thicknesses τ and τ' with $\tau\tau' > 1$, and if neither Cantor set lies in the closure of a component of the complement of the other, then $\Gamma \cap \Gamma'$ must be nonempty. Cantor sets that satisfy these hypotheses are said to be interleaved. Newhouse did not, for his purposes, need to know how big the intersection of two interleaved Cantor sets is. (For an overview of Newhouse'e theory, see [GH, pp. 331–342] or [R, pp. 112–114]. For a more detailed exposition, see [N, pp. 88–109] or [Ro].)

R. Williams in [W] asked about the size of the intersection given by Newhouse's lemma. He gave examples to show that with the hypotheses of Newhouse's lemma, for some pairs of thicknesses the intersection may be just one point. (Once we have defined thickness, we will see that the hypotheses of Newhouse's lemma preclude the trivial case of two Cantor sets that just touch at some common endpoint.) Williams also showed that for some pairs of thicknesses the intersection must contain a Cantor set. Williams posed the problem of determining all pairs of thicknesses for which the intersection of interleaved Cantor sets may be a single point and all pairs of thicknesses for which the intersection of interleaved Cantor sets must contain a Cantor set. This paper gives the solution to this problem. We will also consider the problem of how often, as one Cantor set is being translated over another one, the intersection of the two Cantor sets contains a Cantor set.

The following is an outline of this paper. Section 1 contains the statements of the two main theorems and it also contains six examples which are worked out to provide some motivation for the theorems. The examples demonstrate three basic, nontrivial ways that two Cantor sets can intersect at exactly one point. In Section 2, these three kinds of one point intersections are called overlapped points of the first, second and third kind. In Section 2 we also define the notion of an isolated overlapped point in the intersection of two Cantor sets. We will see that for a pair of thicknesses τ and τ', the question of whether there exist interleaved Cantor sets with these thicknesses that intersect at only one point, or whether the intersection must contain a Cantor set, is equivalent to whether or not there can be an isolated overlapped point in the intersection. Section 2 also contains several lemmas about Cantor sets and thickness that will be used throughout the rest of the paper.

Section 3 contains the proof of Theorem 2 from Section 1, which states that interleaved Cantor sets "almost always" contain a Cantor in their intersection. The rest of the sections make up the proof of Theorem 1 from Section 1, which is the solution to the problem posed

Received by the editors July 2, 1990.

1

by Williams. Section 4 analyzes the case of an overlapped point of the third kind, Section 5 analyzes the case of an overlapped point of the second kind and Sections 6 through 10 analyze the case of an overlapped point of the first kind. In each case, the analysis is done by defining a geometric process that constructs, for any pair of positive numbers (τ, τ') with $\tau\tau' > 1$, two intersecting compact, perfect subsets of the real line. For some choices of (τ, τ'), these sets will be interleaved Cantor sets with thicknesses τ and τ' and with a single overlapped point (of the appropriate kind) in their intersection, and for the other choices of (τ, τ'), these sets will fail to be Cantor sets. We determine for which thicknesses this geometric process succeeds in constructing interleaved Cantor sets, and for which it fails, by translating the geometric process into a two parameter dynamical system (the parameters are τ and τ') and then analyzing the dynamical system. For each kind of overlapped point we get a different dynamical system, and we get different sets of parameter values for which the geometric process succeeds in constructing Cantor sets.

The case of an overlapped point of the first kind is the one that actually gives us the final results as stated in Theorem 1.1. Section 6 defines the geometric process for this case and derives its dynamical system. Section 7 determines the dynamics of this dynamical system. Section 8 uses the dynamics to determine for which thicknesses this geometric process fails and for which it succeeds in constructing Cantor sets. Section 9 then uses these facts about the geometric process to analyze arbitrary Cantor sets with an isolated overlapped point of the first kind. Section 10 proves a part of Theorem 1 from Section 6 about the boundary between two sets in the parameter space of thicknesses. Section 11 completes the proof of Theorem 1.1. Appendix 1 contains the proof of a lemma from Section 7. Appendix 2 contains an alternate proof of Theorem 1.1(2).

The proof of Theorem 1.1 does more than just determine those pairs of thicknesses for which there exist interleaved Cantor sets with one point in their intersection. The proof also gives us information about the possible geometries of the examples that have one point intersections. For example, if $\tau = \tau' = 2$, then there are interleaved Cantor sets Γ and Γ', both with thickness 2, such that $\Gamma \cap \Gamma'$ is a single point. Can Γ and Γ' have as their convex hulls the intervals $[0, 4/5]$ and $[1/5, 1]$? The answer, given by Lemma 9.12, is no. In fact, if $g + g' < 1/2$, then $[0, 1-g]$ and $[g', 1]$ cannot be the convex hulls of Γ and Γ'. Using the information in the proof of Theorem 1.1, it is possible to determine all the possible ways that two interleaved Cantor sets can intersect in just one point.

This paper was my thesis from Northwestern University. I would like to thank my advisor, Clark Robinson, for telling me about this problem and for encouraging me to complete the solution.

1. Theorems and Examples

We will begin by stating the main result of this paper.

Theorem 1.1. *Define sets Λ_1 and Λ_2 by*

$$\Lambda_1 = \left\{ (\tau, \tau') \,\middle|\, \tau > \frac{\tau'^2 + 3\tau' + 1}{\tau'^2} \text{ or } \tau' > \frac{\tau^2 + 3\tau + 1}{\tau^2} \right\}$$

$$\bigcap \left\{ (\tau, \tau') \,\middle|\, \tau > \frac{(1 + 2\tau')^2}{\tau'^3} \text{ or } \tau' > \frac{(1 + 2\tau)^2}{\tau^3} \right\}$$

$$\Lambda_2 = \left\{ (\tau, \tau') \,\middle|\, \tau\tau' > 1 \right\} \setminus \text{closure}(\Lambda_1).$$

Choose τ and τ' such that $\tau\tau' > 1$.

(1) *If $(\tau, \tau') \in \Lambda_1$, then for any pair of interleaved Cantor sets Γ and Γ' with thicknesses τ and τ', $\Gamma \cap \Gamma'$ must contain a Cantor set.*

(2) *If $(\tau, \tau') \in \Lambda_2$, then there exit interleaved, affine Cantor sets Γ and Γ' with thicknesses τ and τ' such that $\Gamma \cap \Gamma'$ is a single point.*

(3) *If (τ, τ') is on the piece of the boundary between Λ_1 and Λ_2 that is defined by the equation*

$$\tau' = \frac{(1 + 2\tau)^2}{\tau^3}$$

with $\tau \in (1 + \sqrt{2}, \tau_0]$, then there exist interleaved Cantor sets Γ and Γ' with thicknesses τ and τ' such that $\Gamma \cap \Gamma'$ is a single point (where τ_0 is the positive real root of $\tau^3 - 3\tau^2 - 4\tau - 1$; τ_0 is approximately equal to 4.04892).

(4) *If $(\tau, \tau') = (1 + \sqrt{2}, 1 + \sqrt{2})$ or if (τ, τ') is on the piece of the boundary between Λ_1 and Λ_2 that is defined by the equation*

$$\tau = \frac{\tau'^2 + 3\tau' + 1}{\tau'^2}$$

with $\tau \in (\tau_0, \infty)$, then for any pair of interleaved Cantor sets Γ and Γ' with thicknesses τ and τ', $\Gamma \cap \Gamma'$ must contain an infinite number of points. However, there exist interleaved, dynamically defined Cantor sets Γ and Γ' with thicknesses τ and τ' such that $\Gamma \cap \Gamma'$ is a countable set.

The definition of an affine Cantor set is given at the end of Section 2. A picture of the sets Λ_1 and Λ_2 is given in Figure 1. Notice that these sets are symmetric with respect to

3

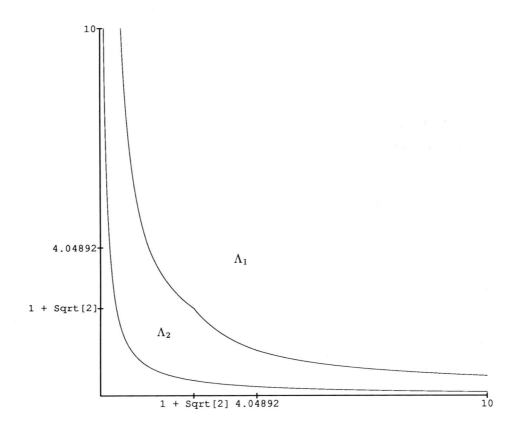

Figure 1

the line $\tau = \tau'$. Parts (3) and (4) of the theorem remain true when the roles of τ and τ' are reversed, so the conclusion of the theorem is symmetric with respect to τ and τ'.

The formulas used in the definition of sets Λ_1 and Λ_2 are derived in this paper in three different ways. The first derivation appears in Example 5 from this section. The most important derivation of these equations is in Section 7, in the proof of Theorem 7.1. And then a third derivation is in Appendix 2. A fourth derivation of these equations can be found in [HKY].

For thicknesses (τ, τ') which are in Λ_1, and for Cantor sets Γ and Γ' with these thicknesses, it is natural to ask for some information about the thickness of the Cantor set that is contained in $\Gamma \cap \Gamma'$. An answer to this question can be found in [HKY].

Let Γ be a Cantor set embedded in the real line. For any real number t, let $\Gamma + t = \{\, x + t \mid x \in \Gamma \,\}$. For any two Cantor sets Γ and Γ' embedded in the real line, let

$$\Gamma - \Gamma' = \{\, t \mid \Gamma \cap (\Gamma' + t) \neq \emptyset \,\}.$$

If we think of the position of Γ as being fixed in the real line and then imagine Γ' as being rigidly translated over Γ, then $\Gamma - \Gamma'$ tells us when the moving copy of Γ' intersects Γ. The set $\Gamma - \Gamma'$ is called the difference set of Γ and Γ' and this name and notation arise because

$$\Gamma - \Gamma' = \{ x - y \mid x \in \Gamma \text{ and } y \in \Gamma' \}.$$

Let

$$T = \{ t \mid \Gamma \text{ and } \Gamma' + t \text{ are interleaved} \}.$$

Suppose that Γ and Γ' have thicknesses τ and τ' and that $\tau\tau' > 1$. Then Newhouse's lemma implies that $T \subset \Gamma - \Gamma'$. In fact, if $\tau\tau' > 1$, then T is all but a countable number of points of $\Gamma - \Gamma'$. If $(\tau, \tau') \in \Lambda_1$, then part (1) of Theorem 1 implies that $\Gamma \cap (\Gamma' + t)$ contains a Cantor set for all $t \in T$. But if $(\tau, \tau') \in \Lambda_2$, then part (2) of Theorem 1 implies that $\Gamma \cap (\Gamma' + t)$ may be as small as a single point for some values of t in T. The question we want to ask is, how often does $\Gamma \cap (\Gamma' + t)$ not contain a Cantor set when $t \in T$? The answer is, not too often. Let

$$T_C = \{ t \mid \Gamma \cap (\Gamma' + t) \text{ contains a Cantor set} \}.$$

In Section 3 we shall prove the following theorem.

Theorem 1.2. *Let Γ and Γ' be Cantor sets embedded in the real line, with thicknesses τ and τ'. If $\tau\tau' > 1$, then T_C is a generic subset of T, in the sense of Baire category.*

Corollary 1.3. *If $\tau\tau' > 1$, then T_C is a generic subset of $\Gamma - \Gamma'$.*

In the rest of this section, we will look at some examples that will help motivate the statements and proofs of the above theorems.

Example 1. In this example, we will construct Cantor sets Γ and Γ' in the intervals $[0, 1]$ and $[0, x]$ for some $x \in (0, 1)$ such that $\Gamma \cap \Gamma'$ will contain only the point 0.

Both Γ and Γ' will be middle-α Cantor sets with the same α for both sets. A middle-α Cantor set (with $\alpha \in (0, 1)$) is defined the same way as the standard middle-1/3 Cantor set is defined, except that we remove from the middle of each closed interval an open interval, that we will call a *gap*, that has length α times the length of the closed interval (instead of an open interval that has length 1/3 that of the closed interval). The two closed subintervals that remain on either side of the gap will be called the *bridges* of the gap. A middle-α Cantor set has thickness

$$\tau = \frac{\text{bridge}}{\text{gap}} = \frac{(1-\alpha)/2}{\alpha} = \frac{1-\alpha}{2\alpha}.$$

Notice that when $\alpha = 1/3$, the thickness is 1. For our purposes, when dealing with middle-α Cantor sets it will be easier to work with $\beta = (1-\alpha)/2$ which is the length of the closed intervals left after the first gap is removed from a middle-α Cantor set in $[0,1]$. In terms of β, the thickness of a middle-α Cantor set is

$$\tau = \frac{\beta}{1-2\beta}.$$

Also notice that $\alpha = 1 - 2\beta$ and the thickness can be written as $\tau = \beta/\alpha$.

Now consider the following picture.

This picture represents the first step in defining two middle-α Cantor sets, one in $[0,1]$ and the other in $[0,x]$. The Cantor set in the interval $[0,x]$ is a rescaled version of the Cantor set in $[0,1]$ (i.e., it is rescaled by the amount x). In order that these Cantor sets have only 0 in their intersection we need $\beta < x(1-\beta)$ and $x < 1-\beta$. In other words, we need

$$\frac{\beta}{1-\beta} < x < 1-\beta.$$

So β must satisfy the inequality $0 < 1 - 3\beta + \beta^2$ which means that

$$\beta < \frac{3 - \sqrt{5}}{2}$$

or

$$\tau < \frac{1 + \sqrt{5}}{2}$$

which is the golden ratio. Notice that when $\tau = (1 + \sqrt{5})/2$, the intersection of the two Cantor sets is the point 0 and also a countable number of endpoints.

Example 2. This will be another example of a way to construct two interleaved Cantor sets that have only one point in their intersection. In this example, the two Cantor sets will again be middle-α, with the same α, but the point that will be in both Cantor sets will not be an endpoint of either Cantor set.

Consider the following picture.

This picture represents the first step in defining two middle-α Cantor sets in the intervals $[0,1]$ and $[x, 1+x]$. The Cantor set in the interval $[x, 1+x]$ is a translate of the Cantor set in $[0, 1]$. In order that these two Cantor sets have only one point in their intersection, we need

$$\beta < x. \tag{1}$$

However, we also need to require that the above picture remains true at all the other stages in the definition of the two Cantor sets. We want to require that the bridge $[1 - \beta, 1]$ from the first Cantor set and the bridge $[x, \beta + x]$ from the second Cantor set be positioned relative to each other the same way that the intervals $[0, 1]$ and $[x, 1+x]$ are positioned. This requirement gives the two Cantor sets together a kind of self-similarity. This requirement can be expressed by

$$\frac{x}{1} = \frac{1 - (\beta + x)}{\beta}$$

or

$$x = \frac{1 - \beta}{1 + \beta}. \tag{2}$$

If we now take the value of x given in equation (2) and put it in the inequality (1), we get

$$\beta < \frac{1 - \beta}{1 + \beta}$$

or

$$\beta^2 + 2\beta - 1 < 0$$

which is solved by

$$\beta < \sqrt{2} - 1$$

which, in terms of thickness, is

$$\tau < 1 + \sqrt{2}.$$

Notice that the upper bound on the thickness in this example is greater than the upper bound in Example 1. Also, when $\beta = \sqrt{2} - 1$ the intersection is a countable number of endpoints along with a unique point that is in the interior of all the overlapping pairs of intervals.

For $\beta \in (0, \sqrt{2} - 1]$, we want to compute what the value is for the unique point that is in the interior of all the overlapping pairs of intervals. The self-similarity of the above construction can be described by a linear map T that maps each Cantor set into itself. This linear map will map the Cantor set in $[0, 1]$ onto its Cantor subset contained in the interval $[1 - \beta, 1]$, and it will map the Cantor set in $[x, 1+x]$ onto its Cantor subset contained in the interval $[x, \beta + x]$. This linear map is defined by the conditions $T(0) = 1$ and $T(1 + x) = x$. The equation for this linear map is

$$T(y) = 1 - \beta y.$$

The unique point that is in the intersection of both Cantor sets must be the unique fixed point of this linear map. So we need to solve $y = 1 - \beta y$, which has the solution

$$y = \frac{1}{1 + \beta}.$$

Notice that when $x = (1-\beta)/(1+\beta)$, the right hand endpoint of the second Cantor set has the value

$$1 + x = 1 + \frac{1-\beta}{1+\beta} = \frac{2}{1+\beta}$$

so the unique intersection point $1/(1+\beta)$ is exactly in the middle of the two extreme endpoints of the two Cantor sets (which is to be expected because of all the symmetry involved). In the next example, we will give another way of computing the value of this intersection point.

Example 3. For any $\beta \in (0, \sqrt{2}-1)$ and any n, this example will modify the last example to get two interleaved Cantor sets whose intersection has 2^n points. Choose $\beta \in (0, \sqrt{2}-1)$ and let

$$x_1 = \beta \left(\frac{1-\beta}{1+\beta} \right)$$

and then consider the following picture.

This picture represents a middle-α Cantor set in the interval $[0,1]$ and a translate of that Cantor set in the interval $[x_1, 1+x_1]$. The intersection of these two Cantor sets will contain two points, which are

$$\frac{\beta}{1+\beta} \quad \text{and} \quad \frac{\beta}{1+\beta} + (1-\beta).$$

In general, for any $\beta \in (0, \sqrt{2}-1)$, let Γ denote the middle-α Cantor set in $[0,1]$ with $\alpha = 1 - 2\beta$. Chose $n \geq 1$ and let

$$x_n = \beta^n \left(\frac{1-\beta}{1+\beta} \right).$$

Let $\Gamma' = \Gamma + x_n$ be the translate of Γ to the interval $[x_n, 1+x_n]$. Then the intersection of Γ and Γ' will contain 2^n points.

Now we will compute the values of these intersection points. For any middle-α Cantor set Γ with fundamental interval $[0,1]$, we can give any point x in Γ a symbolic address of the form $\bar{s} = s_0 s_1 s_2 \ldots s_j \ldots$ where each s_j is equal to either 0 or 1. The value of s_0 is 0 if x is contained in the interval $[0, \beta]$ and it is 1 if x is contained in the interval $[1-\beta, 1]$. In general, the value of s_j is 0 if x is to the left of the nearest gap of length $\alpha\beta^j$ and s_j is 1 if x is to the right of the nearest gap of length $\alpha\beta^j$. Let $x_{\bar{s}}$ denote the point in Γ with address $\bar{s} = s_0 s_1 s_2 \ldots s_j \ldots$. Then the value of $x_{\bar{s}}$ is given by

$$x_{\bar{s}} = \sum_{j=0}^{\infty} s_j \beta^j (1-\beta).$$

In Example 2, the unique intersection point has the following address in the Cantor set Γ,

$$s_0 s_1 s_2 s_3 s_4 s_5 \cdots = 101010 \ldots$$

and the value of this point is

$$\sum_{j=0}^{\infty} s_j \beta^j (1 - \beta) = \sum_{j=0}^{\infty} \beta^{2j} (1 - \beta)$$

$$= (1 - \beta) \sum_{j=0}^{\infty} (\beta^2)^j$$

$$= (1 - \beta) \frac{1}{1 - \beta^2}$$

$$= \frac{1}{1 + \beta}$$

which is the same value that we computed in the last example.

When we chose $n \geq 1$ and $\beta \in (0, \sqrt{2} - 1)$ and let Γ be the middle-α Cantor set with $\alpha = 1 - 2\beta$ and let $\Gamma' = \Gamma + x_n$ with x_n given above, then the 2^n points in the intersection of Γ and Γ' have addresses in Γ of the following form,

$$s_0 s_1 \ldots s_{n-1} 10101010 \ldots$$

where s_j for $0 \leq j \leq n - 1$ can be either 0 or 1. The values of the intersection points with these addresses are given by

$$\frac{\beta^n}{1 + \beta} + \left(\sum_{j=0}^{n-1} s_j \beta^j (1 - \beta) \right).$$

Example 4. For any $\beta \in (0, \sqrt{2} - 1)$ and for any n, this example will modify the last example to give us Cantor sets Γ and Γ', both with thickness $\tau = \beta/(1 - 2\beta)$, such that the intersection of Γ and Γ' will contain n points. The Cantor sets in this example will no longer be middle-α Cantor sets.

First we will do an example that has three intersection points. Fix $\beta \in (0, \sqrt{2} - 1)$ and consider the last example in the case where $n = 1$, so that there are 2^1 intersection points. A picture of the first stage of the construction of the Cantor sets looks like the following picture.

Now, to the right of the intervals marked R and R', place two new intervals, that we will denote R_3 and R_3', so that R_3 and R_3' both have length β and the distance between R

and R_3 is $\alpha = 1 - 2\beta$ and the distance between R' and R'_3 is also α. Then we have the following picture.

$$0 \qquad\qquad\qquad\qquad\qquad R \quad 1 \qquad\qquad R_3$$
$$R' \qquad\qquad\qquad\qquad R'_3$$

Now define middle-α Cantor sets in the intervals R_3 and R'_3. Let Γ be the resulting Cantor set in the interval $[0, 1+\alpha+\beta]$. Let Γ' be the resulting Cantor set in the interval $[x_2, 1+\alpha+\beta+x_2]$. Clearly the intersection of Γ and Γ' contains three points. Using Lemma 2.3 (from the next section) it is easy to prove that Γ and Γ' each have thickness τ.

Now by induction, we can get examples of Cantor sets with any number of intersection points. For example, for $n = 4$, we just add new intervals R_4 and R'_4 to the right of the example that has three intersection points. In general, for $n > 2$ intersection points, the two Cantor sets will be in the intervals $[0, 1+(n-2)(\alpha+\beta)]$ and $[x_2, 1+(n-2)(\alpha+\beta)+x_2]$. The values of the n intersection points are given by

$$\frac{\beta}{1+\beta} + j(1-\beta) \qquad \text{for } j = 0, \ldots, n-1.$$

Example 5. Choose $\beta \in (0, \sqrt{2}-1)$ and let $x = (1-\beta)/(1+\beta)$ as in Example 2. As can be seen in the following picture (this picture has $\beta = 6/17$, so $\tau = 1.2$)

$$0 \qquad\qquad \beta \qquad\quad 1-\beta \qquad\quad 1$$
$$x \qquad\qquad\qquad\qquad\qquad 1+x$$

there is a lot of extra space in the gaps $(\beta, 1-\beta)$ and $(\beta+x, 1-\beta+x)$ which is not occupied by any of the bridges. If we enlarge the bridges into this extra space, then we can increase the thicknesses of these examples. In this example, we will see just how much we can increase these thicknesses.

We will leave the Cantor set in the interval $[x, 1+x]$ unchanged, so it will remain a middle-α Cantor set, and we will increase the thickness of the Cantor set in $[0,1]$ but in such a way that it will no longer be a middle-α Cantor set for any α. In order to see how much we can enlarge the bridges of the Cantor set in $[0,1]$, we need to consider three gaps from the other Cantor set. These gaps have length α/β, length α and length $\beta\alpha$. The following picture shows these three gaps (we get the gap with length $\alpha/\beta = (1-2\beta)/\beta$ by considering the interval $[x, 1+x]$ as the right hand bridge of a middle-α Cantor set in the interval $[1+x-1/\beta, 1+x]$).

$$0 \qquad\quad \beta \quad U \quad 1-\beta \quad 1$$
$$1+x-1/\beta \qquad\qquad x-(1-2\beta)/\beta \qquad x \qquad\qquad 1+x$$

From the picture, we see that the left hand bridge of U can be extended to be the interval $[x-(1-2\beta)/\beta, x]$. The right hand bridge of U can be extended to be the interval $[x+\beta^2, (1+x)-\beta)]$. Then we have the following picture.

Now we have a new gap U' with length $(x + \beta^2) - x = \beta^2$. We have a new left hand bridge of U' with length

$$\frac{1 - 2\beta}{\beta}.$$

And we have a new right hand bridge with length

$$1 - \beta - \beta^2.$$

The ratio determined by the new left hand bridge and the new gap is

$$\frac{(1 - 2\beta)/\beta}{\beta^2} = \frac{1 - 2\beta}{\beta^3}$$

$$= \frac{1 - 2\tau/(1 + 2\tau)}{\tau^3/(1 + 2\tau)^3}$$

$$= \frac{(1 + 2\tau)^2}{\tau^3}.$$

The ratio determined by the new right hand bridge and the new gap is

$$\frac{1 - \beta - \beta^2}{\beta^2} = \frac{1}{\beta^2} - \frac{1}{\beta} - 1$$

$$= \frac{(1 + 2\tau)^2}{\tau^2} - \frac{1 + 2\tau}{\tau} - 1$$

$$= \frac{\tau^2 + 3\tau + 1}{\tau^2}.$$

The thickness determined by the new gap in the interval $[0, 1]$ is equal to

$$\min\left\{ \frac{(1 + 2\tau)^2}{\tau^3}, \frac{\tau^2 + 3\tau + 1}{\tau^2} \right\}$$

where τ, the thickness of the original middle-α Cantor sets, is from the interval $(0, 1 + \sqrt{2})$. As we will show later, the minimum with τ in this interval is given by

$$\frac{\tau^2 + 3\tau + 1}{\tau^2}$$

so this is the largest thickness that the new gap can determine. Since the original construction of the two interleaved Cantor sets from Example 2 was self-similar, the thickening up that we have done at the first level of the Cantor set in $[0, 1]$ can also be done at all the other levels of the original Cantor set. This gives us a Cantor set in $[0, 1]$ that is not middle-α but does have the new thickness

$$\frac{\tau^2 + 3\tau + 1}{\tau^2}$$

and which still has only one point in common with the middle-α Cantor set in $[x, 1 + x]$ (and this common point is the same point as in the construction from Example 2). Notice that these new thicker examples almost account for all of the set Λ_2 from Theorem 1.

Example 6. In this example, we will prove the following theorem about the intersections of a middle-α Cantor set with translates of itself when $\beta = (1 - \alpha)/2$ is sufficiently large.

Theorem 1.4. *If $\beta \in [1/3, 1/2)$ and Γ is the middle-α Cantor set in $[0, 1]$ with $\alpha = 1 - 2\beta$, then there is a set $T \subset [-1, 1]$ such that T has measure 2, T is a generic subset of $[-1, 1]$ in the sense of Baire category, and for all $t \in T$, $\Gamma \cap (\Gamma + t)$ contains a Cantor set.*

Fix $\beta \in (0, 1/2)$ and let Γ denote the middle-α Cantor set in $[0, 1]$ with $\alpha = 1 - 2\beta$. Then

$$\Gamma = \left\{ \sum_{j=0}^{\infty} s_j \beta^j (1 - \beta) \;\middle|\; s_j \in \{0, 1\} \text{ for all } j \right\}.$$

Let $\Gamma_{[-1,1]}$ denote the middle-α Cantor set in $[-1, 1]$. Then

$$\Gamma_{[-1,1]} = \left\{ \sum_{j=0}^{\infty} u_j \beta^j (1 - \beta) \;\middle|\; u_j \in \{-1, 1\} \text{ for all } j \right\}.$$

Let Γ_A denote the union of the images of $\Gamma_{[-1,1]}$ under the affine maps of the form

$$T(x) = \beta^n x + \sum_{j=0}^{n-1} v_j \beta^j (1 - \beta)$$

where $v_j \in \{-1, 0, 1\}$ for $j = 0, 1, 2, \ldots, n - 1$. Then

$$\Gamma_A = \left\{ \sum_{j=0}^{\infty} s_j \beta^j (1 - \beta) \;\middle|\; \begin{array}{l} \text{for some } N, \ s_j \in \{-1, 0, 1\} \text{ for } 0 \leq j \leq N - 1 \\ \text{and } s_j \in \{-1, 1\} \text{ for } j \geq N \end{array} \right\}.$$

Notice that $\Gamma_A \subset [-1, 1]$. Also, Γ_A has measure zero since every middle-α Cantor set has measure zero and hence, Γ_A is a countable union of sets of measure zero. And Γ_A is of first category since every middle-α Cantor set is a nowhere dense set and hence, Γ_A is a countable union of sets of first category.

Lemma 1.5. *Choose $\beta \in (0, 1/2)$. Let Γ, $\Gamma_{[-1,1]}$ and Γ_A be the sets described above. Suppose that $t \notin \Gamma_A$ and that there are points $x, y \in \Gamma$ such that $x - y = t$. Let $\bar{s} = s_0 s_1 s_2 \ldots$ be the address of x in Γ and let $\bar{r} = r_0 r_1 r_2 \ldots$ be the address of y in Γ. Then there is a sequence of integers $\{j_n\}_{n=1}^{\infty}$ such that $s_{j_n} = r_{j_n}$ for $n \geq 1$.*

Proof. Suppose that there is an N such that $s_j \neq r_j$ for all $j \geq N$. Then

$$t = x - y$$

$$= \sum_{j=0}^{\infty} s_j \beta^j (1 - \beta) - \sum_{j=0}^{\infty} r_j \beta^j (1 - \beta)$$

$$= \sum_{j=0}^{N-1} (s_j - r_j) \beta^j (1 - \beta) + \sum_{j=N}^{\infty} (s_j - r_j) \beta^j (1 - \beta)$$

where $s_j - r_j \in \{-1, 0, 1\}$ for $0 \leq j \leq N - 1$ and $s_j - r_j \in \{-1, 1\}$ for $j \geq N$. But then we can see, from the definition of Γ_A above, that $t \in \Gamma_A$, which contradicts one of our hypotheses. So the sequence of integers described in the statement of the lemma must exist. $\square$

Lemma 1.6. *Choose $\beta \in (0, 1/2)$. Let Γ and Γ_A be the sets described above. Suppose that $t \notin \Gamma_A$ and that there are points $x, y \in \Gamma$ such that $x - y = t$. Then $\Gamma \cap (\Gamma + t)$ contains a Cantor set.*

Proof. Let C_0 denote $[0, 1]$ and for $n \geq 1$ let C_n denote the union of the 2^n closed intervals of length β^n that remain after all the gaps from Γ of length greater than or equal to $\alpha \beta^{n-1}$ are removed from $[0, 1]$.

Fix a choice of β from $(0, 1/2)$ and fix a choice of $t \notin \Gamma_A$.

We will show, by induction, that for all $n \geq 0$, there are subsets I_n and I_n' of C_{j_n+1} and $C_{j_n+1} + t$, respectively, such that, each of I_n and I_n' is the union of 2^n closed intervals of length β^{j_n+1}; each component of I_n intersects one component of I_n' (so $I_n \cap I_n'$ contains 2^n closed intervals); and each component of $I_n \cap I_n'$ contains two components of $I_{n+1} \cap I_{n+1}'$. (The sequence of integers $\{j_n\}_{n=1}^{\infty}$ is the sequence given to us by the last lemma, and we will assume that $j_0 = -1$.) Then the set

$$\Lambda = \bigcap_{n=0}^{\infty} (I_n \cap I_n')$$

will be a Cantor set and $\Lambda \subset \Gamma \cap (\Gamma + t)$.

For $n = 0$, let $I_0 = C_0$ and let $I_0' = C_0' + t$. Then $I_0 \cap I_0'$ contains one component.

For $n = 1$, let B_1 and B_1' be the components of C_{j_1} and $C_{j_1} + t$ that contain the point x (so B_1 and B_1' have length β^{j_1}). We know that $s_{j_1} = r_{j_1}$. Suppose, for example, that $s_{j_1} = 0$. Then we have the following picture of B_1 and B_1'.

The two gaps have length $\alpha\beta^{j_1}$ and the four closed intervals have length β^{j_1+1}. (If $s_{j_1} = 1$, then the point x will be to the right of the gaps.) Let I_1 (resp. I_1') denote the two components of C_{j_1+1} (resp. $C_{j_1+1} + t$) pictured above. Then $I_1 \cap I_1'$ has 2 components and both of these components are contained in $I_0 \cap I_0'$.

Now suppose that for some n, we can define the sets I_i and I_i' for $0 \le i \le n$ so that they have the properties described above.

Let B and B' denote the components of I_1 and I_1' respectively that contain the point x (in the picture above, B and B' would be the left hand pair of overlapping intervals). Notice that $\Gamma \cap B$ and $(\Gamma + t) \cap B'$ form a pair of middle-α Cantor sets that satisfy all the hypotheses of this and the previous lemma, but on a different scale and with a new translation value. (If we slide the intervals B and B' over so that B has its left endpoint at 0, and then rescale by the factor $\beta^{-(j_1+1)}$ so that the new intervals have length 1, then the new translation value t_1 is given by $t_1 = \sum_{i=0}^{\infty}(s_{(j_1+1)+i} - r_{(j_1+1)+i})\beta^i(1 - \beta)$ where $s_0 s_1 s_2 \ldots$ and $r_0 r_1 r_2 \ldots$ are the addresses of x and $x - t$ in Γ.) Now apply the induction step to $\Gamma \cap B$ and $(\Gamma + t) \cap B'$ to get sets J_n and J_n' that contain 2^n components of length $\beta^{j(n+1)+1}$ which are all contained in B and B'. Since the right and left halves of the above picture are rigid translations of each other, we can let $I_{n+1} = J_n \cup (J_n \pm \alpha\beta^{j_1})$ and $I_{n+1}' = J_n' \cup (J_n' \pm \alpha\beta^{j_1})$, where we choose the plus sign if $s_{j_1} = 0$ and the minus sign if $s_{j_1} = 1$. $\square$

Theorem 1.7. *If $\beta \in [1/3, 1/2)$ and Γ is the middle-α Cantor set in $[0,1]$ with $\alpha = 1 - 2\beta$, then for all $t \in [-1,1]$, $\Gamma \cap (\Gamma + t) \neq \emptyset$.*

Proof. There are several ways to prove this. The most elegant way is to use a geometric proof like the one found in [B, p.110]. Or one can just apply Newhouse's lemma (Lemma 2.6) since when $\beta \in [1/3, 1/2)$, the thickness of Γ is greater than or equal to one. $\square$

Now we can prove the theorem stated at the beginning of this example, that when $\beta \in [1/3, 1/2)$ and Γ is the middle-α Cantor set in $[0,1]$ with $\alpha = 1 - 2\beta$, then $\Gamma \cap (\Gamma + t)$ contains a Cantor set for "almost all" t in $[-1,1]$ with respect to both Lebesgue measure and Baire category.

Proof. Choose $\beta \in [1/3, 1/2)$. Let Γ_A be defined as above. Choose $t \in [-1,1] \setminus \Gamma_A$. By the last lemma, $\Gamma \cap (\Gamma + t) \neq \emptyset$. So by the second to the last lemma, $\Gamma \cap (\Gamma + t)$ contains a Cantor set. And Γ_A has measure zero and is of first category, so we are done. $\square$

Note. If $\beta \in (0, 1/2)$ and $\Gamma \cap (\Gamma + t) = \{x\}$, then $t \in \Gamma_{[-1,1]}$ (and conversely, if $\beta \in (0, 1/3)$ and $t \in \Gamma_{[-1,1]}$, then $\Gamma \cap (\Gamma + t) = \{x\}$ for some $x \in \Gamma$). Using the language of the next section, if $\beta \in (1/3, 1/2)$ and $\Gamma \cap (\Gamma + t)$ contains an isolated overlapped point, then $t \in \Gamma_A$. So if $t \notin \Gamma_A$, then $\Gamma \cap (\Gamma + t)$ contains no isolated overlapped points and so by Lemma 2.8, $\Gamma \cap (\Gamma + t)$ contains a Cantor set.

2. Cantor Sets and Thickness

In this section we shall gather together some facts about Cantor sets and their thicknesses.

Let Γ be a Cantor set embedded in the real line. Let $\mathcal{U} = \{U_i\}_{i=1}^{\infty}$ be any ordering of the bounded components of $R^1 \setminus \Gamma$. The U_i will be called the *gaps* of Γ. The smallest closed interval I that contains Γ will be called the *fundamental interval* of Γ. When U_1 is removed from I, there remain two subintervals that we will denote $L_1 = L_1(\mathcal{U})$ and $R_1 = R_1(\mathcal{U})$ for left and right subinterval respectively. Let

$$\tau_1(\mathcal{U}) = \min \left\{ \frac{|L_1|}{|U_1|}, \frac{|R_1|}{|U_1|} \right\}$$

where the vertical bars mean the length of the interval. We will call $\tau_1(\mathcal{U})$ the *thickness determined by U_1 in the ordering $\mathcal{U}$*. We will call $L_1(\mathcal{U})$ and $R_1(\mathcal{U})$ the *bridges of U_1 in the ordering $\mathcal{U}$*.

Let $B_1 = B_1(\mathcal{U})$ denote the bridge of U_1 that contains U_2. Then when U_2 is removed from B_1, there remain two subintervals of B_1 which we will denote $L_2 = L_2(\mathcal{U})$ and $R_2 = R_2(\mathcal{U})$. Let

$$\tau_2(\mathcal{U}) = \min \left\{ \frac{|L_2|}{|U_2|}, \frac{|R_2|}{|U_2|} \right\}.$$

In general, let $C_n(\mathcal{U}) = I \setminus \bigcup_{i=1}^{n} U_i$ (so $C_n(\mathcal{U})$ has $n+1$ components). Let B_n be the bridge of C_n that contains U_{n+1}. When U_{n+1} is removed from B_n, the two remaining subintervals of B_n are called $L_{n+1} = L_{n+1}(\mathcal{U})$ and $R_{n+1} = R_{n+1}(\mathcal{U})$. Let

$$\tau_{n+1}(\mathcal{U}) = \min \left\{ \frac{|L_{n+1}|}{|U_{n+1}|}, \frac{|R_{n+1}|}{|U_{n+1}|} \right\}.$$

Let

$$\tau(\mathcal{U}) = \inf\{\tau_n(\mathcal{U})\}_{n=1}^{\infty}.$$

We will call $\tau(\mathcal{U})$ the *thickness determined by the ordering $\mathcal{U}$*.

Notice that given a specific gap U of Γ, the bridges and thickness determined by U depend on the ordering of the gaps as a whole.

The *thickness* of Γ will be defined as

$$\tau = \sup_{\mathcal{U}} \{\tau(\mathcal{U})\}$$

where the supremum is taken over all possible orderings of the bounded components of the complement of Γ.

15

There is a special class of orderings of the gaps of Γ, those where the gaps are ordered by decreasing size (so $|U_{i+1}| \leq |U_i|$). They have the property that they attain the supremum in the definition of thickness.

Theorem 2.1 (Williams [W]). *Let $\mathcal{U} = \{U_i\}_{i=1}^\infty$ be an ordering of the gaps of Γ by decreasing size. Let $\mathcal{V} = \{V_i\}_{i=1}^\infty$ be any other ordering of the gaps. Then $\tau(\mathcal{V}) \leq \tau(\mathcal{U})$.*

Proof. We will show that given any m, there is an n such that

$$\tau_n(\mathcal{V}) \leq \tau_m(\mathcal{U}),$$

which implies that $\inf\{\tau_i(\mathcal{V})\} \leq \inf\{\tau_i(\mathcal{U})\}$.

Suppose that $R_m(\mathcal{U})$ is the bridge of U_m that determines $\tau_m(\mathcal{U})$ (there is a similar argument when $L_m(\mathcal{U})$ determines $\tau_m(\mathcal{U})$). If the right hand endpoint of $R_m(\mathcal{U})$ is also the right hand endpoint of the fundamental interval I of Γ, then the right hand bridge of U_m in the $\mathcal{V}$ ordering can be no longer then $R_m(\mathcal{U})$. So in this case, let n be such that V_n is the same gap as U_m. Then

$$\tau_n(\mathcal{V}) \leq \frac{|R_n(\mathcal{V})|}{|V_n(\mathcal{V})|} = \frac{|R_n(\mathcal{V})|}{|U_m(\mathcal{U})|} \leq \frac{|R_m(\mathcal{U})|}{|U_m(\mathcal{U})|} = \tau_m(\mathcal{U}).$$

If the right hand endpoints of $R_m(\mathcal{U})$ and I are not equal, then there is a gap U_j on the right end of $R_m(\mathcal{U})$ with $j < m$, and so $|U_j| \geq |U_m|$. In the ordering $\mathcal{V}$, the gap U_j can come before or after the gap U_m. Suppose that in $\mathcal{V}$, U_j comes before U_m. Then let n be such that V_n is the gap U_m. Then $|R_n(\mathcal{V})| \leq |R_m(\mathcal{U})|$ and so we again have

$$\tau_n(\mathcal{V}) \leq \frac{|R_n(\mathcal{V})|}{|V_n(\mathcal{V})|} = \frac{|R_n(\mathcal{V})|}{|U_m(\mathcal{U})|} \leq \frac{|R_m(\mathcal{U})|}{|U_m(\mathcal{U})|} = \tau_m(\mathcal{U}).$$

Now suppose that in $\mathcal{V}$, U_j comes after U_m. Then let n be such that V_n is the gap U_j. Then $|L_n(\mathcal{V})| \leq |R_m(\mathcal{U})|$ and

$$\tau_n(\mathcal{V}) \leq \frac{|L_n(\mathcal{V})|}{|V_n(\mathcal{V})|} \leq \frac{|R_m(\mathcal{U})|}{|U_j(\mathcal{U})|} \leq \frac{|R_m(\mathcal{U})|}{|U_m(\mathcal{U})|} = \tau_m(\mathcal{U}). \qquad \square$$

A useful consequence of this theorem is the following lemma.

Lemma 2.2. *Let Γ have thickness τ and let $\mathcal{U} = \{U_i\}_{i=1}^\infty$ be an ordering of the gaps of Γ by decreasing size. Then for any i, the Cantor sets $R_i(\mathcal{U}) \cap \Gamma$ and $L_i(\mathcal{U}) \cap \Gamma$ each have thickness greater than or equal to τ.*

Proof. Choose either R_i or L_i. Let $\mathcal{V} \subset \mathcal{U}$ be the gaps of Γ that are contained in the chosen bridge. Then $\inf\{\tau_i(\mathcal{V})\} \geq \inf\{\tau_i(\mathcal{U})\} = \tau$. $\qquad \square$

The last lemma is not true for arbitrary orderings of the gaps of Γ. For example, let Γ be the middle-third Cantor set in $[0, 1]$. Let $U_1 = (1/9, 2/9)$ and $U_2 = (1/3, 2/3)$. Then Γ has thickness 1 but $R_1 \cap \Gamma$ has thickness $\frac{1/9}{1/3} = 1/3$.

Another useful lemma (which does not depend on Theorem 2.1) is

Lemma 2.3. *Suppose $\mathcal{V} = \{V_i\}_{i=1}^{\infty}$ is an ordering of the gaps of a Cantor set Γ such that there is a number τ with*

$$\tau_i(\mathcal{V}) = \tau \quad \text{for all } i.$$

Then the thickness of Γ is τ.

Proof. Let $\mathcal{U} = \{U_i\}_{i=1}^{\infty}$ be any ordering of the gaps of Γ (not necessarily by decreasing size). We will show that $\tau(\mathcal{U}) \leq \tau(\mathcal{V})$.

Suppose V_1 and U_1 are the same gap. Then $\tau_1(\mathcal{U}) = \tau$ so $\tau(\mathcal{U}) \leq \tau = \tau(\mathcal{V})$.

Now suppose $V_1 \neq U_1$. Let m be such that U_m is the same gap as V_1. Then $m > 1$. But U_m in the $\mathcal{U}$ ordering will have shorter bridges then V_1 in the $\mathcal{V}$ ordering. So $\tau_m(\mathcal{U}) \leq \tau_1(\mathcal{V}) = \tau$. Therefore $\tau(\mathcal{U}) \leq \tau = \tau(\mathcal{V})$. $\qquad\square$

A generalization of this last lemma, which has the exact same proof, is

Lemma 2.4. *Suppose $\mathcal{V} = \{V_i\}_{i=1}^{\infty}$ is an ordering of the gaps of a Cantor set Γ and $\mathcal{V}$ has the property that*

$$\tau_i(\mathcal{V}) \geq \tau_1(\mathcal{V}) \quad \text{for all } i.$$

Then the thickness of Γ is $\tau_1(\mathcal{V})$.

If $\mathcal{U} = \{U_i\}_{i=1}^{\infty}$ is an ordering by decreasing size of the gaps of the Cantor set Γ, then $\tau_i(\mathcal{U}) \geq \tau$ for all i, where τ is the thickness of Γ. Sometimes we will need to work with subsequences of $\mathcal{U}$. If $\mathcal{V}$ is an arbitrary subsequence of $\mathcal{U}$, then we want to prove that $\mathcal{V}$ keeps the property that $\tau_i(\mathcal{V}) \geq \tau$ for all i. However, if we remove only the gaps in $\mathcal{V}$ from the fundamental interval of Γ, then the bridges $L_i(\mathcal{V})$ and $R_i(\mathcal{V})$ need not be fundamental intervals of Cantor sets.

Lemma 2.5. *If $\mathcal{V} = \{V_i\}_{i=1}^{\infty}$ is a subsequence of all the gaps of a Cantor set Γ and if $\mathcal{V}$ is ordered by decreasing size, then $\tau_i(\mathcal{V}) \geq \tau$ for all i, where τ is the thickness of Γ.*

Proof. Any gap V_i in the ordering $\mathcal{V}$ will have bridges that are longer then its bridges in any ordering $\mathcal{U}$ of all the gaps of Γ by decreasing size of which $\mathcal{V}$ is a subsequence. So $\tau_i(\mathcal{V}) \geq \tau_i(\mathcal{U}) \geq \tau$ for all i. $\qquad\square$

We will define two Cantor sets Γ and Γ' to be *interleaved* if neither one lies in the closure of a gap of the other. If B is a bridge of Γ (for some ordering of the gaps of Γ) and B' is a bridge of Γ', then we say that B and B' are *interleaved bridges* if $B \cap \Gamma$ and $B' \cap \Gamma'$ are interleaved Cantor sets.

Suppose Γ and Γ' are Cantor sets such that $\Gamma \cap \Gamma' \neq \emptyset$. Let $\mathcal{U} = \{U_i\}_{i=1}^{\infty}$ and $\mathcal{U}' = \{U_i'\}_{i=1}^{\infty}$ be orderings by decreasing size of the gaps of Γ and Γ'. If $x \in \Gamma \cap \Gamma'$, then for every n, $x \in C_n(\mathcal{U}) \cap C_n(\mathcal{U}')$. Let A_n be the bridge of $C_n(\mathcal{U})$ that contains x and let A_n' be the bridge of $C_n(\mathcal{U}')$ that contains x. So $x \in A_n \cap A_n'$ for all n. We will say that x is *overlapped* if one of the following three cases holds:

(1) $x \in \text{interior}(A_n)$ and $x \in \text{interior}(A'_n)$ for all n,

(2) $x \in \text{interior}(A_n)$ for all n and there is an n such that x is an endpoint of A'_n or $x \in \text{interior}(A'_n)$ for all n and there is an n such that x is an endpoint of A_n,

(3) there is an n such that x is an endpoint of both A_n and A'_n and $A_n \cap A'_n \neq \{x\}$.

The following three pictures give an idea of what the three different kinds of overlapped points look like in the bridges A_n and A'_n. The first kind looks like this.

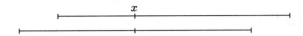

The second kind looks like this.

And the third kind looks like this.

The point $x \in \Gamma \cap \Gamma'$ will not be overlapped if there is an n such that A_n and A'_n look like this.

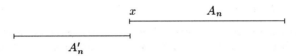

Now we can state and prove Newhouse's lemma. The proof given here is a variation on the proof given in [GH, p. 334].

Lemma 2.6 (Newhouse). *Let Γ and Γ' be Cantor sets with thicknesses τ and τ' such that $\tau\tau' > 1$. If Γ and Γ' are interleaved, then there is a point x such that $x \in \Gamma \cap \Gamma'$ and x is overlapped.*

Note. We can replace the hypothesis $\tau\tau' > 1$ with $\tau\tau' \geq 1$ and still conclude that $\Gamma \cap \Gamma' \neq \emptyset$, but we would not be able to guarantee the existence of an overlapped point in the intersection.

Proof. Let $\mathcal{U} = \{U_i\}_{i=1}^{\infty}$ and $\mathcal{U}' = \{U'_i\}_{i=1}^{\infty}$ be orderings, by decreasing size, of the gaps of Γ and Γ' respectively. We will show, by induction, that for all n, $C_n(\mathcal{U}) \cap C_n(\mathcal{U}')$ has nonempty interior.

Suppose that $C_1(\mathcal{U}) \cap C_1(\mathcal{U}')$ has empty interior. Since Γ and Γ' are interleaved, this implies that we have a situation that looks like the following picture.

But then we have

$$\tau\tau' \leq \frac{|L_1|}{|U_1|}\frac{|R_1'|}{|U_1'|} \leq \frac{|U_1'|}{|U_1|}\frac{|U_1|}{|U_1'|} = 1$$

which contradicts the hypothesis that $\tau\tau' > 1$.

Now assume that $C_{n-1}(\mathcal{U}) \cap C_{n-1}(\mathcal{U}')$ has nonempty interior.

Consider all the pairs of components (B, B') from $C_{n-1}(\mathcal{U})$ and $C_{n-1}(\mathcal{U}')$ such that $B \cap B'$ has nonempty interior.

Suppose there is a pair of these components that looks like this

that is, there is a pair of components that satisfy

$$B \setminus (B \cap B') \neq \emptyset \qquad \text{and} \qquad B' \setminus (B \cap B') \neq \emptyset. \tag{$*$}$$

If $U_n \not\subset B$ or if $U_n' \not\subset B'$, then clearly $C_n(\mathcal{U}) \cap C_n(\mathcal{U}')$ has nonempty interior. On the other hand, if $U_n \subset B$ and $U_n' \subset B'$ but $(B \setminus U_n) \cap (B' \setminus U_n')$ has empty interior, then we have a picture exactly like the one above, with just the labels being changed:

which, by Lemma 2.2, contradicts the hypothesis that $\tau\tau' > 1$. So in this case, $(B \setminus U_n) \cap (B' \setminus U_n')$ must have nonempty interior, which also implies that $C_n(\mathcal{U}) \cap C_n(\mathcal{U}')$ has nonempty interior.

Now suppose that condition $(*)$ is not satisfied by any pair of components (B, B') from $C_{n-1}(\mathcal{U})$ and $C_{n-1}(\mathcal{U}')$. So if $B \cap B'$ has nonempty interior, then either $B \subset B'$ or $B' \subset B$. Suppose that $C_n(\mathcal{U}) \cap C_n(\mathcal{U}')$ has empty interior. Then either the closure of U_n contains components of $C_n(\mathcal{U}')$ or the closure of U_n' contains components of $C_n(\mathcal{U})$. We will assume the former case, that the closure U_n contains components of $C_n(\mathcal{U}')$, and we will also assume that the closure of U_n contains only one component B' of $C_n(\mathcal{U}')$ (the case where the closure of U_n contains several components of $C_n(\mathcal{U}')$ is similar). Let V' denote the shorter of the two gaps on either side of B' in $C_n(\mathcal{U}')$ (one of these gaps must have finite length). Suppose that V' is on the right of B'. Then we have the following picture.

Notice that $|R_n| \leq |V'|$ since we are assuming that $C_n(\mathcal{U}) \cap C_n(\mathcal{U}')$ has empty interior. So now we have

$$\tau\tau' \leq \frac{|R_n|}{|U_n|}\frac{|B'|}{|V'|} \leq \frac{|V'|}{|U_n|}\frac{|U_n|}{|V'|} = 1$$

which, because of Lemma 2.2, contradicts the hypothesis that $\tau\tau' > 1$. So $C_n(\mathcal{U}) \cap C_n(\mathcal{U}')$ must have nonempty interior.

So for all n, $C_n(\mathcal{U}) \cap C_n(\mathcal{U}')$ has nonempty interior and therefore $\Gamma \cap \Gamma' \neq \emptyset$.

Suppose that all the points in $\Gamma \cap \Gamma'$ are nonoverlapped. By the definition of nonoverlapped, it is clear that a nonoverlapped point is isolated in $\Gamma \cap \Gamma'$. So by the compactness of $\Gamma \cap \Gamma'$, if all its point are nonoverlapped, there can only be a finite number of them. But then there is some sufficiently large N such that all the bridges of $C_N(\mathcal{U})$ are in the closure of gaps of $C_N(\mathcal{U}')$ and visa-versa, which contradicts that $C_N(\mathcal{U}) \cap C_N(\mathcal{U}')$ has nonempty interior. So $\Gamma \cap \Gamma'$ must contain an overlapped point.   □

Define an *isolated overlapped point* in the intersection of two Cantor sets to be an overlapped point that has an open neighborhood about it that does not contain any other overlapped points (but this neighborhood may contain nonoverlapped points from the intersection). The following lemma is a version of a lemma that Williams in [W] called the Cantor Lemma. So we will give it the same name here.

Lemma 2.7(Cantor Lemma). *Suppose that Γ and Γ' are interleaved Cantor sets with thicknesses τ and τ' where $\tau\tau' > 1$. Also suppose that $\Gamma \cap \Gamma'$ does not contain an isolated overlapped point. Let $\mathcal{U} = \{U_i\}_{i=1}^{\infty}$ and $\mathcal{U}' = \{U_i'\}_{i=1}^{\infty}$ be orderings of the gaps of Γ and Γ' by decreasing size. For any n, if B and B' are interleaved bridges of $C_n(\mathcal{U})$ and $C_n(\mathcal{U}')$, then there is an $m > n$ such that $B \cap C_m(\mathcal{U})$ and $B' \cap C_m(\mathcal{U}')$ contain two pairs of interleaved bridges.*

Proof. Choose a pair of interleaved bridges B and B' from $C_n(\mathcal{U})$ and $C_n(\mathcal{U}')$ for some n. By Newhouse's lemma, there must be an overlapped point x in $B \cap B'$. Suppose that x is an overlapped point of the first kind. Choose an ϵ small enough so that $(x - \epsilon, x + \epsilon) \subset B \cap B'$. Since x is not an isolated overlapped point, there is another overlapped point in this ϵ-neighborhood of x. Let y denote this other overlapped point. Then x and y are each contained in B and B'. Now choose $m > n$ large enough so that the bridge of $C_m(\mathcal{U})$ that contains x, call it B_1, is disjoint from the bridge of $C_m(\mathcal{U})$ that contains y, which we shall call B_2, and also so that the bridge of $C_m(\mathcal{U}')$ that contains x, call it B_1', is disjoint from the bridge of $C_m(\mathcal{U}')$ that contains y, which we shall call B_2'. Because x and y are each contained in B, both B_1 and B_2 are subsets of B. Similarly, since x and y are each contained in B', both B_1' and B_2' are subsets of B'. Since x is an overlapped point and x is in both B_1 and B_1', these two bridges are interleaved. Since y is overlapped and in both B_2 and B_2', these two bridges are interleaved. So we have shown that $B \cap C_m(\mathcal{U})$ and $B' \cap C_m(\mathcal{U}')$ contain two pairs of interleaved bridges.

The cases where x is an overlapped point of the second or third kind are similar.   □

The Cantor Lemma allows us to prove that no isolated overlapped points in the intersection of interleaved Cantor sets implies that the intersection contains a Cantor set.

Lemma 2.8. *Suppose that Γ and Γ' are interleaved Cantor sets with thicknesses τ and τ', where $\tau\tau' > 1$, such that $\Gamma \cap \Gamma'$ does not contain an isolated overlapped point. Then the intersection of Γ and Γ' must contain a Cantor set.*

Proof. Let I_0 be the intersection of the fundamental intervals of Γ and Γ'. By the previous lemma, there is an n_1 such that $C_{n_1}(\mathcal{U})$ and $C_{n_1}(\mathcal{U}')$ have two pair of interleaved bridges. Let I_1 be the intersections of these two pairs of bridges. Then $I_1 \subset I_0$ and I_1 has two components. Using the previous lemma again, we get an $n_2 > n_1$ such that the last two pairs of interleaved bridges each contain two pairs of interleaved bridges. Let I_2 be the intersection of these four pairs of interleaved bridges. Then $I_2 \subset I_1$ and I_2 contains four components. Continuing by induction, we get an increasing sequence of integers n_j, such that $C_{n_j}(\mathcal{U})$ and $C_{n_j}(\mathcal{U}')$ have 2^j pairs of interleaved bridges. Each of the 2^{j-1} pairs of interleaved bridges from $C_{n_{j-1}}(\mathcal{U})$ and $C_{n_{j-1}}(\mathcal{U}')$ contain two of the 2^j pairs of bridges from $C_{n_j}(\mathcal{U})$ and $C_{n_j}(\mathcal{U}')$. So we get a sequence of nested sets I_j made up of the intersections of the 2^j pairs of bridges in $C_{n_j}(\mathcal{U})$ and $C_{n_j}(\mathcal{U}')$. Each I_j has 2^j components.

Now we shall show that

$$\Lambda = \bigcap_{j=0}^{\infty} I_j$$

is a Cantor set. It is clear that Λ is a compact set since it is the intersection of compact sets. We need to show that Λ is totally disconnected and perfect. Since each component of I_j is the intersection of a bridge from $C_{n_j}(\mathcal{U})$ and a bridge from $C_{n_j}(\mathcal{U}')$, and since the lengths of these bridges go to zero as $j \to \infty$, the lengths of the components of I_j must also go to zero as $j \to \infty$. This implies that Λ cannot contain any intervals. So Λ is totally disconnected. To show that Λ is a perfect set, choose an x in Λ and choose $\epsilon > 0$. We need to find a y in Λ such that $|x - y| < \epsilon$. Choose a j large enough so that all the bridges in $C_{n_j}(\mathcal{U})$ and $C_{n_j}(\mathcal{U}')$ have length strictly less than ϵ. Let B and B' denote the bridges of $C_{n_j}(\mathcal{U})$ and $C_{n_j}(\mathcal{U}')$ that contain x. By the way we defined n_{j+1}, $C_{n_{j+1}}(\mathcal{U})$ and $C_{n_{j+1}}(\mathcal{U}')$ will contain two pairs of overlapped bridges that are subsets of B and B'. One of these pairs will contain x. The other pair will contain a point of Λ that cannot be x, so call this point y. Then $|x - y| < \epsilon$ because both x and y are contained in B which has length strictly less than ϵ. So Λ is a perfect set.

Finally, it is easy to show that $\Lambda \subset \Gamma \cap \Gamma'$. $\square$

Note. The existence of a nonisolated overlapped point in $\Gamma \cap \Gamma'$ is not sufficient to imply that the intersection contains a Cantor set. Moreover, it is also not necessary that every overlapped point in the intersection be nonisolated for the intersection to contain a Cantor set.

The simple idea of an isolated overlapped point can be used to restate the problem posed by Williams and described in the introduction. For any pair of thicknesses (τ, τ') with $\tau\tau' > 1$, we ask if we can find two Cantor sets Γ and Γ' with these thicknesses such that $\Gamma \cap \Gamma'$ contains an isolated overlapped point. If no such pair of Cantor sets can be found, then by the last lemma, $\Gamma \cap \Gamma'$ must contain a Cantor set for any two interleaved Cantor sets with thicknesses τ and τ'. On the other hand, if two Cantor sets exist such that their intersection contains an isolated overlapped point, then it is not hard to show that we can also find Cantor sets Γ and Γ' with thicknesses τ and τ' such that $\Gamma \cap \Gamma'$ contains only one overlapped point (though it may also contain up to a countable number of nonoverlapped points).

Now we want to look more carefully at the idea of choosing a number τ and then removing a gap from a closed interval $[a, b]$ such that the thickness determined by the gap is exactly τ.

Suppose we are given $\tau > 0$, an interval $I = [a, b]$ and a number $\theta \in (0, 1)$. We want to remove a gap U from I such that the thickness determined by U in I is τ and so that the center of U is at the point $(1 - \theta)a + \theta b$.

$$
\begin{array}{cccc}
\overset{\displaystyle L}{\vdash\!\!\!\!\!\!-\!\!\!-\!\!\!-\!\!\!-\!\!\!\dashv} & \overset{\displaystyle U}{\vdash\!\dashv} & & \overset{\displaystyle R}{\vdash\!\!-\!\!\!-\!\!\!-\!\!\!-\!\!\!-\!\!\!-\!\!\!-\!\!\!-\!\!\!-\!\!\!\dashv} \\
a & (1 - \theta)a + \theta b & & b
\end{array}
$$

Let L and R be the left and right hand bridges of U in I. If $\theta \leq 1/2$, then L will be shorter than R and so L will determine the thickness of U in I. If $\theta \geq 1/2$, then R will be shorter and R will determine the thickness.

Suppose $\theta \leq 1/2$. We shall derive formulas for $|L|$, $|R|$ and $|U|$ as functions of τ and θ. We start with $|L| = \theta(b - a) - |U|/2$. Then since $\theta \leq 1/2$,

$$
\tau = \frac{|L|}{|U|} = \frac{\theta(b - a) - |U|/2}{|U|}
$$

so

$$
|U| = \frac{2\theta}{1 + 2\tau}(b - a). \tag{1}
$$

And since $|L| = |U|\tau$, we get

$$
|L| = \frac{2\tau\theta}{1 + 2\tau}(b - a).
$$

And finally

$$
\begin{aligned}
|R| &= (b - a) - |U| - |L| \\
&= (b - a) - \frac{2\theta}{1 + 2\tau}(b - a) - \frac{2\tau\theta}{1 + 2\tau}(b - a) \\
&= \left[1 - \frac{2\theta(1 + \tau)}{1 + 2\tau}\right](b - a).
\end{aligned}
$$

If $\theta > 1/2$, then the formulas for $|L|$ and $|R|$ are exchanged and θ is replaced by $1 - \theta$.

These formulas show how a gap can "slide" across the interval I as θ slides across the interval $(0, 1)$ while keeping constant the thickness that the gap determines. Notice that as the gap U slides across I from left to right, $|L|$ is strictly increasing and $|R|$ is strictly decreasing. Also, the ratio $|L|/|U|$ is increasing while the ratio $|R|/|U|$ is decreasing as θ goes from 0 to 1.

Also notice that when $\theta = 1/2$, $|L| = |R| = \dfrac{\tau}{1 + 2\tau}(b - a)$. In the following sections, we will use the following simple fact many times.

Lemma 2.9. *Suppose that an open interval U is removed from a closed interval $[a, b]$ and that U determines thickness τ in $[a, b]$. If B is a bridge of U and if*

$$|B| \leq \frac{\tau}{1 + 2\tau}(b - a),$$

then B is the bridge of U that determines the thickness.

We can now use the above ideas to describe a generalization of the construction of a middle-α Cantor set that will be called a θ-α Cantor set, where $\theta \in (0, 1)$ and $\alpha \in (0, 1)$. In a middle-α Cantor set, gaps are always removed from the middle of the closed intervals. So for all of these gaps, the value of θ that determines the location of the center of the gap is always equal to $1/2$. To construct a θ-α Cantor set, gaps are removed from closed intervals $[a, b]$ so that their center is at the point $(1 - \theta)a + \theta b$ and so that the thickness determined by the gap is $(1 - \alpha)/2\alpha$. In other words, we start with the same gap as for a middle-α construction and then slide that gap over so that it has a new center but so that it still determines the same thickness as it did in the middle-α construction. As in the construction of a middle-α Cantor set, we begin with a fundamental interval I, then remove the first (and largest) gap, which leaves two bridges. Then the same construction is repeated on each bridge. By Lemma 3, the thickness of a θ-α Cantor set will be $(1-\alpha)/2\alpha$, the same thickness as for the middle-α Cantor set. Notice that a middle-α Cantor set is the same as a $1/2$-α Cantor set.

A nice application of θ-α Cantors sets is the following. Let $\mathrm{HD}(\Gamma)$ denote the Hausdorff dimension of a Cantor set Γ. In [PT] it is shown that if τ is the thickness of Γ, then we have the following lower bound on the Hausdorff dimension of Γ

$$\mathrm{HD}(\Gamma) \geq \frac{\log 2}{-\log(\tau/(1 + 2\tau))}.$$

If Γ is a middle-α Cantor set, it is well known that

$$\mathrm{HD}(\Gamma) = \frac{\log 2}{-\log \beta} = \frac{\log 2}{-\log(\tau/(1 + 2\tau))}.$$

So a middle-α Cantor set has the minimum Hausdorff dimension possible for a given thickness. Now if we fix α and let θ vary from $1/2$ to 1, then it is not hard to see that the Hausdorff dimension of the θ-α Cantor sets will vary from the lower bound up to 1. So for fixed α, the one parameter family of θ-α Cantor sets have fixed thickness but all possible Hausdorff dimensions.

The last thing we will do in this section is to give the definition of an affine Cantor set. This definition of an affine Cantor set is taken from [PT]. An affine Cantor set is a special case of what Palis and Takens in [PT] call a dynamically defined Cantor set.

Let $\{J_j\}_{j=1}^k$ be a finite collection of closed subintervals of the interval $[0, 1]$. Let $J_j = [a_j, b_j]$. Suppose that these subintervals satisfy

(1) $a_1 = 0$,
(2) $a_j < b_j$ for $1 \leq j \leq k$,
(3) $b_j < a_{j+1}$ for $1 \leq j \leq k - 1$,
(4) $b_k = 1$.

Now define k affine maps by the formulas

$$T_j(x) = (b_j - a_j)x + a_j.$$

Notice that $T_j(0) = a_j$ and $T_j(1) = b_j$ so T_j maps $[0,1]$ onto $[a_j, b_j]$. Now let $I_0 = [0,1]$ and inductively define

$$I_{n+1} = \bigcup_{j=1}^{k} T_j(I_n).$$

Notice that $I_1 = \cup_{j=1}^{k} J_j$. Also notice that I_n has k^n components, that $I_{n+1} \subset I_n$, and the lengths of the components of I_n go to zero as $n \to \infty$ (because the affine maps T_j are contractions). Now let

$$\Gamma = \bigcap_{n=0}^{\infty} I_n.$$

Then Γ is an affine Cantor set. We will call Γ the affine Cantor set in $[0,1]$ with template $\{J_j\}_{j=1}^{k}$.

An arbitrary Cantor set Γ contained in $[0,1]$ is said to be an affine Cantor set if we can find a template $\{J_j\}_{j=1}^{k}$ such that Γ equals the Cantor set defined by this template using the construction described above. Given an affine Cantor set Γ, a template for Γ is not unique. For example, given any template $\{J_j\}_{j=1}^{k}$ for Γ, any one of the collection of intervals I_n defined above is also a template for Γ (but with k^n affine maps).

Notice that any θ-α-Cantor set is an affine Cantor set. Also, the definition of an affine Cantor set can be extended, in an obvious way, to the case where the fundamental interval I_0 is not $[0,1]$.

3. Proof of Theorem 1.2

In the this section we will prove Theorem 1.2. First we need to prove a few lemmas.

Lemma 3.1. *If Γ and Γ' are interleaved Cantor sets, then there is an $\epsilon > 0$ such that if $|t| < \epsilon$, then Γ and $\Gamma' + t$ are also interleaved. So being interleaved is an open condition and the set $T = \{\, t \mid \Gamma$ and $\Gamma' + t$ are interleaved $\}$ is an open set.*

Proof. Let I and I' be the fundamental intervals of Γ and Γ'. We need to consider three cases. In the first case, one endpoint of I' is in the interior of I and the other endpoint of I' is not in I.

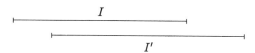

In this case, it is clear that the conclusion of the lemma is true (let ϵ be the minimum distance between an endpoint of I and an endpoint of I').

In the second case, I and I' have a common endpoint. Assume that $|I| \geq |I'|$. Let U be a gap of Γ such that the bridge from the common endpoint of I and I' to the nearest endpoint of U is strictly shorter than I'. Denote this bridge by B. Then we have the following picture.

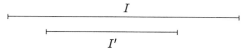

From the picture, it is clear that for sufficiently small $t < 0$, the intervals I and $I' + t$ are interleaved. For sufficiently small $t > 0$, the intervals B and $I' + t$ are interleaved. Let ϵ be the smaller of these two conditions.

In the third case, both endpoints of one fundamental interval are contained in the interior of the other fundamental interval. We will assume that $I' \subset I$.

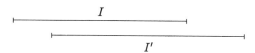

Let U be a gap of I that intersects I'. Since Γ and Γ' are interleaved, I' cannot be contained in the closure of U. So part of I' is outside the closure of U. Then from the next picture, it is easy to see that one of the bridges between U and the endpoints of I is interleaved with I' in the manner of the first case above.

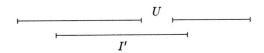

Lemma 3.2. *Suppose that* Γ *and* Γ' *are Cantor sets with thicknesses* τ *and* τ' *with* $\tau\tau' > 1$. *Then there exists a* t *such that the intersection of* Γ *and* $\Gamma' + t$ *contains two overlapped points.*

Proof. Let I be the fundamental interval of Γ and let I' be the fundamental interval of Γ'. Let $\mathcal{U} = \{U_i\}_{i=1}^{\infty}$ be an ordering of the gaps of Γ by decreasing size. Choose n such that $|U_n| < |I'|$. Choose t_1 such that the left hand endpoint of $I' + t_1$ is equal to the left hand endpoint of U_n. Then we need to consider the following two cases.

The first case is if the right hand endpoint $I' + t_1$ is less than or equal to the right hand endpoint of R_n. The second case is if the right hand endpoint of $I' + t_1$ is strictly greater than the right hand endpoint of R_n.

In the first case, it is clear that R_n and $I' + t_1$ are interleaved bridges. By the previous lemma, being interleaved is an open condition. So we can find an $\epsilon > 0$ such that L_n and $I' + (t_1 - \epsilon)$ are interleaved and such that R_n and $I' + (t_1 - \epsilon)$ are still interleaved. By Lemma 2.2 and Newhouse's lemma, each one of these pairs of interleaved bridges will give us an overlapped point. So let $t = t_1 - \epsilon$.

In the second case, notice that R_n cannot be contained in the closure of a gap of $I' + t_1$ because of our hypothesis that $\tau\tau' > 1$. Let $U' + t_1$ be the largest gap of $\Gamma' + t_1$ that intersects R_n.

Then R_n must be interleaved with one of the bridges of $U' + t_1$. Since being interleaved is an open condition, we can find an $\epsilon > 0$ such that L_n is interleaved with $I' + (t_1 - \epsilon)$ and R_n is still interleaved with the same bridge of $U' + (t_1 - \epsilon)$. By Lemma 2.2 and Newhouse's lemma, each one of these pairs of interleaved bridges will give us an overlapped point. So let $t = t_1 - \epsilon$.   □

Now we will prove the following Cantor like lemma.

Lemma 3.3. *Suppose that* Γ *and* Γ' *are interleaved Cantor sets with thicknesses* τ *and* τ' *with* $\tau\tau' > 1$. *Then, for any* $\epsilon > 0$ *there is a* t *such that* $|t| < \epsilon$ *and the Cantor sets* Γ *and* $\Gamma' + t$ *have two overlapped points in their intersection.*

Proof. By Newhouse's lemma, Γ and Γ' must contain an overlapped point x in their intersection. We will prove that for any $\epsilon > 0$ there is a t such that $|t| < \epsilon$, the Cantor sets Γ

and $\Gamma' + t$ have two overlapped points in their intersection, and that the two overlapped points are within a distance $\epsilon/2$ of x (so in a sense, the overlapped point x splits into two overlapped points).

Let $\mathcal{U} = \{U_i\}_{i=1}^{\infty}$ and $\mathcal{U}' = \{U_i'\}_{i=1}^{\infty}$ be orderings of the gaps of Γ and Γ' by decreasing size. Choose a value of n large enough so that the bridges of C_n and C_n' that contain x have length strictly less than $\epsilon/2$. Let B and B' denote these bridges. By the last lemma, there is a t such that the Cantor sets $B \cap \Gamma$ and $(B' \cap \Gamma') + t$ have two overlapped points in their intersection.

Claim. $|t| < \epsilon$.

When $t = 0$, B and B' intersect so $|B \cup B'| < \epsilon$. If $|t| \geq \epsilon$, either $t > 0$ and the left hand endpoint of $B' + t$ is to the right of the right hand endpoint of B, or $t < 0$ and the right hand endpoint of $B' + t$ is to the left of the left hand endpoint of B. In both of these cases, B and $B' + t$ do not intersect, which contradicts that they contain overlapped points. So the claim must be true.

Since x and both of the overlapped points contained in Γ and $\Gamma' + t$ are contained in the bridge B, which has length strictly less than $\epsilon/2$, the new overlapped points are within a distance $\epsilon/2$ of x.   □

Let Γ and Γ' be Cantor sets with thicknesses τ and τ' with $\tau\tau' > 1$. Let

$$T = \{\, t \mid \Gamma \text{ and } \Gamma' + t \text{ are interleaved} \,\}$$

and let

$$T_C = \{\, t \mid \Gamma \cap (\Gamma' + t) \text{ contains a Cantor set} \,\}.$$

Theorem 3.4. *T_C is dense in T. In particular, given any overlapped point x in the intersection of the Cantor sets Γ and Γ', then for all $\epsilon > 0$ there is a t with $|t| < \epsilon$ such that $\Gamma \cap (\Gamma' + t)$ contains a Cantor set which lies within a distance ϵ of x.*

Note. This theorem shows that any overlapped point, and in particular an isolated overlapped point, can "explode" into a Cantor set for arbitrarily small translations of one of the Cantor sets, when the product of the thicknesses is strictly greater than one.

Proof. Let $t \in T$ and choose $\epsilon > 0$. We need to find a t_0 such that $|t - t_0| < \epsilon$ and $t_0 \in T_C$. Without lose of generality, we may assume that $t = 0$. Let $\mathcal{U} = \{U_i\}_{i=1}^{\infty}$ and $\mathcal{U}' = \{U_i'\}_{i=1}^{\infty}$ be orderings of the gaps of Γ and Γ' by decreasing size. Since Γ and Γ' are interleaved, there is an overlapped point x in their intersection. By the last lemma, we can find a t_1 such that $|t_1| < \epsilon/2$ and such that there are two pairs of interleaved bridges for Γ and $\Gamma' + t_1$ within $\epsilon/2$ of x. Denote one pair of these bridges by B_1 and $B_1' + t_1$ and the other pair by B_2 and $B_2' + t_1$. Since being interleaved is an open condition, there is a $\delta_1 > 0$ and a $\delta_1' > 0$ such that $|s| < \min\{\delta_1, \delta_1'\}$ implies that Γ and $(\Gamma' + t_1) + s$ still have these same pairs of interleaved bridges. Using the last lemma, choose a t_2 such that

$$|t_2| < \min\{\delta_1, \delta_1', \epsilon/4\}$$

and such that the pair of interleaved bridges B_1 and $B_1' + t_1 + t_2$ split into two pairs of interleaved bridges that we will denote B_{11}, $B_{11}' + t_1 + t_2$ and B_{12}, $B_{12}' + t_1 + t_2$. Using the last lemma again, choose a t_3 such that $|t_3| < \epsilon/8$ and so that the pairs B_{11}, $B_{11}' + t_1 + t_2 + t_3$ and B_{12}, $B_{12}' + t_1 + t_2 + t_3$ both stay interleaved but the pair B_2 and $B_2' + t_1 + t_2 + t_3$ splits into two pairs of interleaved bridges that we will denote B_{21}, $B_{21}' + t_1 + t_2 + t_3$ and B_{22}, $B_{22}' + t_1 + t_2 + t_3$. We now have a situation for the Cantor sets Γ and $\Gamma' + (t_1 + t_2 + t_3)$ that looks something like the following picture (where $t' = t_1 + t_2 + t_3$).

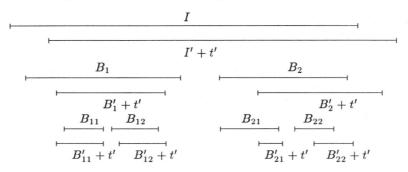

After four more applications of the last lemma, one for each of the interleaved pairs above, we will have eight pairs of interleaved bridges, that we can denote $B_{i_1 i_2 i_3}$ and $B_{i_1 i_2 i_3}' + (t_1 + t_2 + t_3 + t_4 + t_5 + t_6 + t_7)$ with each i_j being 1 or 2, for the Cantor sets Γ and $\Gamma' + (t_1 + t_2 + t_3 + t_4 + t_5 + t_6 + t_7)$ where each t_i satisfies $|t_i| < \epsilon/2^i$.

If we keep this construction going on for ever, we will get a sequence $\{t_i\}_{i=1}^{\infty}$ such that $|t_i| < \epsilon/2^i$. Then the infinite series

$$t_0 = \sum_{i=1}^{\infty} t_i$$

converges, and its sum t_0 is strictly less than ϵ. The Cantor sets Γ and $\Gamma' + t_0$ will have the 2^n nested, interleaved pairs of bridges $B_{i_1...i_n}$ and $B_{i_1...i_n}' + t_0$ for all n. So the intersection of Γ and $\Gamma' + t_0$ will contain a Cantor set which is within a distance ϵ of the overlapped point x. So T_C is dense in T.   □

Now we can prove a weak version of Theorem 1.2.

Theorem 3.5. *Let*

$$T_I = \{\, t \mid \Gamma \cap (\Gamma' + t) \text{ has infinite cardinality}\,\}.$$

Then T_I is a generic subset of T (in the sense of Baire category).

Proof. Let

$$T_n = \{\, t \mid \Gamma \cap (\Gamma' + t) \text{ has at least } 2^n \text{ overlapped points}\,\}.$$

Notice that $T_0 = T$. In the proof of the last theorem, we showed that each T_n is an open and dense subset of the open set T. So the set

$$T_\infty = \bigcap_{n=0}^{\infty} T_n$$

is a generic subset of T. And clearly, $T_\infty \subset T_I$.   □

Another description of each T_n is given by

$$T_n = \{\, t \mid \Gamma \cap (\Gamma' + t) \text{ has at least } 2^n \text{ pairs of interleaved bridges}\,\}.$$

It would be nice to conclude that $T_\infty \subset T_C$. When $t \in T_\infty$, we know that for any n we can find a pair of interleaved bridges B_n and B'_n that split into 2^n pairs of interleaved bridges with the property that each of the pairs of bridges at a level $j < n$ splits into two pairs of bridges at the level $j + 1$. But when we look at the pair of bridges B_{n+1} and B'_{n+1} that split into 2^{n+1} pairs of interleaved bridges, we do not know that $B_n = B_{n+1}$ and that $B'_n = B'_{n+1}$. So it may not be true that we can find one particular pair of interleaved bridges B and B' that keeps on splitting forever in the way that is needed to guarantee a Cantor set in the intersection of Γ and $\Gamma' + t$.

Let

$$T_E = \{\, t \in T \mid \text{ no endpoint of } \Gamma \text{ is equal to an endpoint of } \Gamma' + t\,\}.$$

Since the complement of T_E in T is a countable set, T_E is a generic subset of T. Let

$$T_{1/n} = \{\, t \in T_E \mid \text{ every overlapped point of } \Gamma \cap (\Gamma' + t)$$
$$\text{has another overlapped point within } 1/n \text{ of it}\,\}.$$

Lemma 3.6. *For all n, $T_{1/n}$ is an open and dense subset of T_E.*

The proof of this lemma will be broken up into two separate lemmas.

Lemma 3.7. *For all n, $T_{1/n}$ is an open subset of T_E.*

Proof. Let $\mathcal{U} = \{U_i\}_{i=1}^\infty$ and $\mathcal{U}' = \{U'_i\}_{i=1}^\infty$ be orderings of the gaps of Γ and Γ' by decreasing size. Fix a value of n and choose an m sufficiently large so that all the bridges of $C_m(\mathcal{U})$ and $C_m(\mathcal{U}')$ have length strictly less than $1/n$. Choose $t \in T_{1/n}$. We need to find an ϵ so that if $|s - t| < \epsilon$ and if $s \in T_E$, then $s \in T_{1/n}$.

Let B and $B' + t$ be any two bridges from $C_m(\mathcal{U})$ and $C_m(\mathcal{U}') + t$. Either these two bridges are disjoint or they are not. Suppose that B and $B' + t$ are disjoint. Then we can find an $\epsilon_{BB'}$ such that if $|s - t| < \epsilon_{BB'}$, then B and $B' + s$ are still disjoint.

Suppose that B and $B' + t$ are not disjoint. Then either they are interleaved or they are not interleaved. Suppose that B and $B' + t$ are not interleaved. Then one of these bridges must lie in the closure of a gap from the other Cantor set. But since $t \in T_E$, we can in fact conclude that one of these bridges lies in a gap of the other Cantor set. Assume, for example, that B lies in a gap of $(\Gamma' \cap B') + t$. Since gaps are open intervals, there is an $\epsilon_{BB'}$ such that if $|s - t| < \epsilon_{BB'}$ then B still lies in a gap of $(\Gamma' \cap B') + s$.

Suppose that B and $B' + t$ are interleaved. Then either the intersection of Γ and $\Gamma' + t$ contains only one overlapped point in $B \cap (B' + t)$ or the intersection of Γ and $\Gamma' + t$ contains at least two overlapped points in $B \cap (B' + t)$. Suppose that there are at least two overlapped points in $B \cap (B' + t)$. That means that the bridges B and $B' + t$ eventually split into two pairs of interleaved subbridges. Denote one of these pairs by B_1 and $B'_1 + t$ and the other pair by B_2 and $B'_2 + t$. Since being interleaved is an open condition (Lemma 1) there is an $\epsilon_{BB'}$ such that if $|s - t| < \epsilon_{BB'}$, then B_1 and $B'_1 + s$ are still interleaved and B_2 and $B'_2 + s$ are still interleaved. So, in particular, the bridges B and $B' + s$ still contain at least two overlapped points when $|s - t| < \epsilon_{BB'}$.

Suppose that B and $B' + t$ contain only one overlapped point. Let x denote this overlapped point. Since $t \in T_{1/n}$, there must be another overlapped point y in $\Gamma \cap (\Gamma' + t)$ such that $y \neq x$ and $|y - x| < 1/n$. Let $\delta = 1/n - |y - x|$. Choose an integer k such that $k > m$ and all the bridges of $C_k(\mathcal{U})$ and $C_k(\mathcal{U}')$ have length strictly less than $\delta/2$. Let B_1 and $B_1' + t$ denote the pair of interleaved bridges from $C_k(\mathcal{U})$ and $C_k(\mathcal{U}') + t$ that contain x. Let A and $A' + t$ denote the pair of interleaved bridges from $C_k(\mathcal{U})$ and $C_k(\mathcal{U}') + t$ that contain y. Note that

$$\max\{\, |u - v| \mid u \in B_1 \text{ and } v \in A \,\} < 1/n.$$

Since being interleaved is an open condition, there is an $\epsilon_{BB'}$ such that if $|s - t| < \epsilon_{BB'}$, then the pair of bridges B_1 and $B_1' + s$ are still interleaved and the pair A and $A' + s$ are also still interleaved. So in particular, if $|s - t| < \epsilon_{BB'}$, any overlapped point in $B \cap (B' + s)$ still has another overlapped point in $A \cap (A' + s)$ within distance $1/n$.

So far we have shown that if we take any pair of bridges B and $B' + t$ from $C_m(\mathcal{U})$ and $C_m(\mathcal{U}') + t$, we can find an $\epsilon_{BB'}$ such that if $|s - t| < \epsilon_{BB'}$, and if there are no overlapped points in $B \cap (B' + t)$, then there are no overlapped points in $B \cap (B' + s)$, and on the other hand, if there is an overlapped point in $B \cap (B' + t)$, then every overlapped point in $B \cap (B' + s)$ has another distinct overlapped point from $\Gamma \cap (\Gamma' + s)$ within $1/n$. Now let

$$\epsilon = \min\{\, \epsilon_{BB'} \mid B \text{ and } B' + t \text{ are bridges from } C_m(\mathcal{U}) \text{ and } C_m(\mathcal{U}') + t \,\}.$$

Then if $|s - t| < \epsilon$ and if y is an overlapped point of $\Gamma \cap (\Gamma' + s)$, then either y must be in a bridge of $C_m(\mathcal{U})$ that has two overlapped points from $\Gamma \cap (\Gamma' + t)$ and therefore, by the choice of ϵ, also has two overlapped points from $\Gamma \cap (\Gamma' + s)$, or y is in a bridge of $C_m(\mathcal{U})$ that has only one overlapped point from $\Gamma \cap (\Gamma' + t)$ but by the choice of ϵ again, there must still be another overlapped point from $\Gamma \cap (\Gamma' + s)$ within $1/n$ of y. So then $(t - \epsilon, t + \epsilon) \cap T_E$ is an open neighborhood of t in $T_{1/n}$. So each $T_{1/n}$ is open in T_E. □

Note. The set

$$\{\, t \in T \mid \text{every overlapped point of } \Gamma \cap (\Gamma' + t)$$
$$\text{has another overlapped point within } 1/n \text{ of it} \,\}$$

is not an open subset of T. The problem is that if for some t there is a nonoverlapped point x in $\Gamma \cap (\Gamma' + t)$, then for s arbitrarily close to t it may be that $\Gamma \cap (\Gamma' + s)$ has an isolated overlapped point y, near where x was, and such that y has no other overlapped points within $1/n$ of it. For example, consider the Cantor sets in Example 3 from Section 1. Our hypothesis that $t \in T_E$ eliminates the possibility of nonoverlapped points in the intersection of Γ and $\Gamma' + t$.

Lemma 3.8. *For all n, $T_{1/n}$ is a dense subset of T_E.*

Proof. Fix a value of n. Choose $t \in T_E$. Choose $\epsilon > 0$. We will assume that $\epsilon < 1/n$. We need to find a t_0 in $T_{1/n}$ such that $|t_0 - t| < \epsilon$.

Let $\mathcal{U} = \{U_i\}_{i=1}^{\infty}$ and $\mathcal{U}' = \{U_i'\}_{i=1}^{\infty}$ be orderings of the gaps of Γ and Γ' by decreasing size. Choose an m sufficiently large so that all the bridges of $C_m(\mathcal{U})$ and $C_m(\mathcal{U}')$ have length strictly less than $1/n$. Let B and $B' + t$ be any pair of bridges from $C_m(\mathcal{U})$ and $C_m(\mathcal{U}') + t$. Then we have one of two cases for B and $B' + t$, either they are interleaved or they are not interleaved. Choose ϵ_1 such that, for any pair of interleaved bridges B and $B' + t$ from $C_m(\mathcal{U})$ and $C_m(\mathcal{U}') + t$, if $|s - t| < \epsilon_1$, then B and $B' + s$ are also interleaved. Choose ϵ_2 such that, for any pair of noninterleaved bridges B and $B' + t$ from $C_m(\mathcal{U})$ and $C_m(\mathcal{U}') + t$, if $|s - t| < \epsilon_2$, then B and $B' + s$ are also noninterleaved (such an ϵ_2 exits because of our hypothesis that $t \in T_E$). Let

$$\epsilon_3 = \min\{\epsilon, \epsilon_1, \epsilon_2\}.$$

Now consider only those pairs of bridges B and $B' + t$ from $C_m(\mathcal{U})$ and $C_m(\mathcal{U}') + t$ that are interleaved. There is some finite number k of such pairs. Give these k pairs some ordering and denote them by B_i and $B'_i + t$ for $i = 1 \ldots k$. By Lemma 3, there is an s_1 contained in the ϵ_3 neighborhood of t such that B_1 and $B'_1 + s_1$ contain two overlapped points. By Lemma 1, there is an open neighborhood of s_1 such that B_1 and $B'_1 + s$ contain two overlapped points for any s in this neighborhood. Assume that this neighborhood of s_1 is contained in the ϵ_3 neighborhood of t. Using Lemma 3 again, we can find an s_2 contained in this neighborhood of s_1 such that B_2 and $B'_2 + s_2$ contain two overlapped points. And we can find a neighborhood of s_2 such that B_2 and $B'_2 + s$ contain two overlapped points for any s in this neighborhood of s_2. Assume that this neighborhood of s_2 is contained in the neighborhood of s_1. By finite induction up to k, we can find an open set $\mathcal{O}$ contained in the open interval $(t - \epsilon_3, t + \epsilon_3)$ such that for all $s \in \mathcal{O}$, each of the interleaved pairs of bridges from $C_m(\mathcal{U})$ and $C_m(\mathcal{U}') + s$ contain at least two overlapped points. Now let $t_0 \in \mathcal{O} \cap T_E$ and we are done. $\quad\square$

Let

$$T_{1/\infty} = \bigcap_{n=1}^{\infty} T_{1/n}.$$

Lemma 3.9. $T_{1/\infty}$ *is a generic subset of* T.

Proof. We will show that $T_{1/\infty}$ can be written as the countable intersection of open and dense subsets of T. Since the complement of T_E in T is countable, we can write T_E as the countable intersection of open and dense subsets G_n of T. Since each $T_{1/n}$ is open in T_E, there exist open subsets O_n of T such that $T_{1/n} = T_E \cap O_n$. Since each $T_{1/n}$ is dense in T_E, and since T_E is dense in T, each O_n is dense in T. Then

$$T_{1/\infty} = \bigcap_{n=1}^{\infty} T_{1/n}$$

$$= \bigcap_{n=1}^{\infty} (T_E \cap O_n)$$

$$= T_E \cap \left(\bigcap_{n=1}^{\infty} O_n \right)$$

$$= \left(\bigcap_{n=1}^{\infty} G_n \right) \cap \left(\bigcap_{n=1}^{\infty} O_n \right)$$

$$= \bigcap_{n=1}^{\infty} (G_n \cap O_n). \qquad \square$$

The next lemma completes the proof of Theorem 1.2.

Lemma 3.10. $T_{1/\infty} \subset T_C$.

Proof. Let $t \in T_{1/\infty}$. Then $t \in T_{1/n}$ for all n. So if x is an overlapped point of $\Gamma \cap (\Gamma' + t)$, there are other overlapped points $y_n \neq x$ of $\Gamma \cap (\Gamma' + t)$ within $1/n$ of x for all n. So $\Gamma \cap (\Gamma' + t)$ does not contain any isolated overlapped points. Then, by Lemma 2.8, the intersection of Γ and $\Gamma' + t$ must contain a Cantor set. So $t \in T_C$. $\qquad \square$

4. Third Kind Of Overlapped Point

In this section we will consider the possibility of two Cantor sets Γ and Γ' with $\Gamma \cap \Gamma' = \{x\}$ where x is an overlapped point of the third kind. We will assume that the Cantor sets Γ and Γ' have fundamental intervals $I = [0,1]$ and $I' = [0,1-g]$ respectively where $g \in (0,1)$ and we will determine for which thicknesses it is possible that $\Gamma \cap \Gamma' = \{0\}$. For those thicknesses for which it is not possible that $\Gamma \cap \Gamma' = \{0\}$, we will prove a partial version of the Cantor Lemma.

The results of this section are stated in the following theorem.

Theorem 4.1. *Define sets Λ_3 and Λ_4 by*

$$\Lambda_3 = \left\{ (\tau, \tau') \,\Big|\, \tau > \frac{1 + 2\tau'}{\tau'^2} \text{ or } \tau' > \frac{1 + 2\tau}{\tau^2} \right\}$$

$$\Lambda_4 = \{ (\tau, \tau') \mid \tau\tau' > 1 \} \setminus \text{closure}(\Lambda_3)$$

Let Γ and Γ' be Cantor sets with fundamental intervals $I = [0,1]$ and $I' = [0,1-g]$ with $g \in (0,1)$ and let τ and τ' be their thicknesses.

(1) *If $(\tau, \tau') \in \Lambda_3$, then $\Gamma \cap \Gamma' \neq \{0\}$. In fact, there is an n such that C_n and C_n' have two pairs of interleaved bridges.*

(2) *If $(\tau, \tau') \in \Lambda_4$, then it may be that $\Gamma \cap \Gamma' = \{0\}$.*

(3) *If (τ, τ') is on the boundary between Λ_3 and Λ_4, then $\Gamma \cap \Gamma'$ must have infinite cardinality but it may be that the cardinality is countably infinite and 0 is the only overlapped point.*

This theorem will be proven by defining an iterated geometric process that will attempt to construct Cantor sets Γ and Γ' in the intervals I and I' with thicknesses τ and τ' and such that $\Gamma \cap \Gamma' = \{0\}$. The geometric process will succeed for thicknesses $(\tau, \tau') \in \Lambda_4$. For thicknesses $(\tau, \tau') \in \Lambda_3$, the process will fail to define Cantor sets.

The geometric process will be analyzed be translating it into a function $\Phi_{\tau\tau'}$ that can be iterated and then analyzing the resulting dynamical system. The thicknesses τ and τ' are parameters for the dynamical system and the set Λ_4 will be seen to be those parameter values for which the dynamical system has nonzero fixed points.

To define the geometric process, we start with the intervals $I = [0,1]$ and $I' = [0,1-g]$ with $g \in (0,1)$ and with two positive numbers τ, τ' with $\tau\tau' > 1$. The first step of the first iteration of the process is to remove a gap U_1 from I such that the right hand endpoint of U_1 is the point $1-g$ and so that U_1 determines thickness τ in I. The second step of the first iteration is to remove a gap U_1' from the interval I' so that the right hand endpoint of U_1' is equal to the left hand endpoint of U_1 and so that U_1' determines thickness τ' in I'. At this point we have the following picture.

33

The diagram at top shows intervals labeled 0, L_1, U_1, R_1, 1 on the first line and L_1', U_1', R_1', $1-g$ on the second line.

The last step of the first iteration is to construct a middle-α Cantor set in bridge R_1 with thickness τ and a middle-α Cantor set with thickness τ' in bridge R_1'. These three steps make up one iterate of the geometric process.

Notice that the bridges L_1 and L_1' can be considered as new fundamental intervals, analogous to I and I', so the process can be iterated over again, this time applied to L_1 and L_1'. Clearly, this process can be iterated an infinite number of times.

The diagram shows intervals labeled 0, L_3, 1 on the first line and L_3', $1-g$ on the second line.

This process will define two Cantor sets if and only if, as $n \to \infty$, the lengths of the bridges L_n and L_n' go to zero. To see that this may not always be the case, take for example τ and τ' very large and g very small. It seems intuitively plausible that the lengths of the gaps U_n and U_n' will go to zero so fast that the lengths of L_n and L_n' will remain bounded away from zero. See the following picture.

The diagram shows intervals labeled 0, 1 on the first line and $1-g$ on the second line.

When the process does define two Cantor sets, they are obviously interleaved and their thicknesses will be τ and τ' by Lemma 2.3. The Cantor sets will have as their intersection the point zero and a countable number of endpoints of bridges. The endpoints can be removed by making arbitrarily small changes in the thicknesses τ and τ', as will be shown latter (this is Williams' "shaving principle" [W]).

To be able to analyze this geometric process, we will now translate it into a function that can be iterated. First find equations for $|U_1|$ and $|U_1'|$. Since

$$\tau = \min\left\{\frac{|L_1|}{|U_1|}, \frac{|R_1|}{|U_1|}\right\}$$

we have

$$|U_1| = \begin{cases} \dfrac{|R_1|}{\tau} & \text{if } |R_1| \leq \dfrac{\tau}{1+2\tau} \\[2mm] \dfrac{|L_1|}{\tau} & \text{if } |R_1| \geq \dfrac{\tau}{1+2\tau} \end{cases}$$

$$= \begin{cases} \dfrac{g}{\tau} & \text{if } g \leq \dfrac{\tau}{1+2\tau} \\[2mm] \dfrac{1-g-|U_1|}{\tau} & \text{if } g \geq \dfrac{\tau}{1+2\tau}. \end{cases}$$

Solving for $|U_1|$ in the second part of the equation we get

$$|U_1| = \begin{cases} \dfrac{g}{\tau} & \text{if } g \leq \dfrac{\tau}{1+2\tau} \\ \dfrac{1-g}{1+\tau} & \text{if } g \geq \dfrac{\tau}{1+2\tau}. \end{cases}$$

In a similar way we also get

$$|U_1'| = \begin{cases} \dfrac{|R_1'|}{\tau'} & \text{if } |R_1'| \leq \dfrac{\tau'}{1+2\tau'}(1-g) \\ \dfrac{|L_1'|}{\tau'} & \text{if } |R_1'| \geq \dfrac{\tau'}{1+2\tau'}(1-g) \end{cases}$$

$$= \begin{cases} \dfrac{|U_1|}{\tau'} & \text{if } |U_1| \leq \dfrac{\tau'}{1+2\tau'}(1-g) \\ \dfrac{1-g-|U_1|-|U_1'|}{\tau'} & \text{if } |U_1| \geq \dfrac{\tau'}{1+2\tau'}(1-g). \end{cases}$$

By solving for $|U_1'|$ in the second part of the equation, we get

$$|U_1'| = \begin{cases} \dfrac{|U_1|}{\tau'} & \text{if } |U_1| \leq \dfrac{\tau'}{1+2\tau'}(1-g) \\ \dfrac{1-g-|U_1|}{1+\tau'} & \text{if } |U_1| \geq \dfrac{\tau'}{1+2\tau'}(1-g). \end{cases}$$

The inequalities in the above equations were arrived at by using Lemma 2.9.

After we have removed the gaps U_1 and U_1', we need, for the purpose of the function that we wish to define, to rescale the bridges L_1 and L_1' so that L_1 will have length 1. Then the distance from the right hand endpoint of the rescaled L_1' (i.e., from the point $|L_1'|/|L_1|$) to the point 1 will be a new value of g which we will denote $\Phi_{\tau\tau'}(g)$. So we define the function $\Phi_{\tau\tau'}$ by

$$\Phi_{\tau\tau'}(g) = 1 - \frac{|L_1'|}{|L_1|}.$$

Notice that $|L_1| = 1 - g - |U_1|$ and $|U_1'| = |L_1| - |L_1'|$ so

$$\Phi_{\tau\tau'}(g) = 1 - \frac{|L_1'|}{|L_1|} = \frac{|L_1| - |L_1'|}{|L_1|} = \frac{|U_1'|}{1-g-|U_1|}.$$

Suppose that $g \leq \tau/(1+2\tau)$. Then $|U_1| = g/\tau$ so

$$|U_1'| = \begin{cases} \dfrac{g}{\tau\tau'} & \text{if } g/\tau \leq \dfrac{\tau'}{1+2\tau'}(1-g) \\ \dfrac{1-g-g/\tau}{1+\tau'} & \text{if } g/\tau \geq \dfrac{\tau'}{1+2\tau'}(1-g). \end{cases}$$

Then

$$\Phi_{\tau\tau'}(g) = \frac{|U_1'|}{1 - g - |U_1|}$$

$$= \begin{cases} \dfrac{g}{\tau\tau'(1 - g - g/\tau)} & \text{if } g \le \dfrac{\tau\tau'}{\tau\tau' + 2\tau' + 1} \\[3ex] \dfrac{1}{1 + \tau'} & \text{if } g \ge \dfrac{\tau\tau'}{\tau\tau' + 2\tau' + 1}. \end{cases}$$

At this point, it is worth noting that

$$\frac{\tau}{1 + 2\tau} \le \frac{\tau\tau'}{\tau\tau' + 2\tau' + 1} \quad \text{if and only if} \quad \tau\tau' - \tau' - 1 \ge 0.$$

Now suppose the other case, that $g \ge \tau/(1 + 2\tau)$. Then $|U_1| = (1 - g)/(1 + \tau)$ so

$$\Phi_{\tau\tau'}(g) = \frac{|U_1'|}{1 - g - |U_1|}$$

$$= \begin{cases} \dfrac{1 - g}{\tau'(1 + \tau)\left(1 - g - \dfrac{1 - g}{1 + \tau}\right)} & \text{if } \dfrac{1 - g}{1 + \tau} \le \dfrac{\tau'}{1 + 2\tau'}(1 - g) \\[5ex] \dfrac{1 - g - \dfrac{1 - g}{1 + \tau}}{(1 + \tau')\left(1 - g - \dfrac{1 - g}{1 + \tau}\right)} & \text{if } \dfrac{1 - g}{1 + \tau} \ge \dfrac{\tau'}{1 + 2\tau'}(1 - g) \end{cases}$$

$$= \begin{cases} \dfrac{1}{\tau\tau'} & \text{if } 0 \le \tau\tau' - \tau' - 1 \\[2ex] \dfrac{1}{1 + \tau'} & \text{if } 0 \ge \tau\tau' - \tau' - 1. \end{cases}$$

So the definition of $\Phi_{\tau\tau'}$ has two cases depending on the sign of $\tau\tau' - \tau' - 1$.

Proposition 4.2.

If $\tau\tau' - \tau' - 1 \ge 0$, then

$$\Phi_{\tau\tau'}(g) = \begin{cases} \dfrac{g}{\tau\tau'(1 - g - g/\tau)} & \text{if } g \le \dfrac{\tau}{1 + 2\tau} \\[3ex] \dfrac{1}{\tau\tau'} & \text{if } g \ge \dfrac{\tau}{1 + 2\tau}. \end{cases}$$

If $\tau\tau' - \tau' - 1 \le 0$, then

$$\Phi_{\tau\tau'}(g) = \begin{cases} \dfrac{g}{\tau\tau'(1 - g - g/\tau)} & \text{if } g \le \dfrac{\tau\tau'}{\tau\tau' + 2\tau' + 1} \\[3ex] \dfrac{1}{1 + \tau'} & \text{if } \dfrac{\tau\tau'}{\tau\tau' + \tau' + 1} \le g \le \dfrac{\tau}{1 + 2\tau} \\[3ex] \dfrac{1}{1 + \tau'} & \text{if } g \ge \dfrac{\tau}{1 + 2\tau}. \end{cases}$$

Notice that
$$\frac{1}{\tau\tau'} = \frac{1}{1+\tau'} \quad \text{if and only if} \quad \tau\tau' - \tau' - 1 = 0.$$

Now that we have defined a two parameter family of maps $\Phi_{\tau\tau'}$, we need to analyze the dynamics for all parameter values τ and τ' with $\tau\tau' > 1$.

Note. From now on, the domain of $\Phi_{\tau\tau'}$ will always be $[0,1)$, even though $g = 0$ does not make any sense in the geometric model that $\Phi_{\tau\tau'}$ was derived from. Also, $\Phi_{\tau\tau'}$ is continuous but not differentiable.

The dynamics of $\Phi_{\tau\tau'}$ are described by the following proposition.

Proposition 4.3. *Assume that* $\tau\tau' > 1$.

(1) *Zero is a hyperbolic sink for* $\Phi_{\tau\tau'}$.
(2) *If* τ *and* τ' *also satisfy the inequalities*

$$\tau' \le \frac{1+2\tau}{\tau^2} \quad \text{and} \quad \tau \le \frac{1+2\tau'}{\tau'^2}$$

then the point
$$g = \frac{\tau\tau' - 1}{\tau'(1+\tau)}$$

is a fixed point of $\Phi_{\tau\tau'}$ *and also either*

$$g = \frac{1}{\tau\tau'} \quad \text{or} \quad g = \frac{1}{1+\tau'}$$

is a fixed point of $\Phi_{\tau\tau'}$ *(depending on the sign of* $\tau\tau' - \tau' - 1$*).*
(3) *If* τ *and* τ' *do not satisfy the inequalities in the last item, then zero is the only fixed point of* $\Phi_{\tau\tau'}$ *and its basin of attraction is all of* $[0,1)$.

Note. If τ and τ' satisfy strict inequalities in item (2), then the first fixed point listed is repelling and the other fixed point is attracting. When (τ, τ') crosses from the set Λ_3 to the set Λ_4 the map $\Phi_{\tau\tau'}$ undergoes a nondifferentiable saddle-node bifurcation.

The proof of this proposition is given by the following four lemmas.

Lemma 4.4. *If* $\tau\tau' > 1$*, then zero is a hyperbolic sink for* $\Phi_{\tau\tau'}$.

Proof. Clearly $\Phi_{\tau\tau'}(0) = 0$ for all τ, τ'. Also

$$d\Phi_{\tau\tau'}\Big|_{g=0} = \frac{1}{\tau\tau'} \frac{1 - g\left(\frac{1+\tau}{\tau}\right) + g\left(\frac{1+\tau}{\tau}\right)}{\left(1 - g\left(\frac{1+\tau}{\tau}\right)\right)^2}\Bigg|_{g=0} = \frac{1}{\tau\tau'} < 1.$$

$\square$

Let $\overline{\Phi}_{\tau\tau'}$ denote the function

$$\overline{\Phi}_{\tau\tau'}(g) = \frac{g}{\tau\tau'\left(1 - g - g/\tau\right)}.$$

The domain of $\overline{\Phi}_{\tau\tau'}$ will be $[0, \tau/(1+\tau))$ and the range of $\overline{\Phi}_{\tau\tau'}$ is $[0, \infty)$.

Besides zero, another fixed point of $\overline{\Phi}_{\tau\tau'}$ can be found by solving

$$g = \frac{g}{\tau\tau'\left(1 - g - g/\tau\right)}$$

which has the solution

$$g = \frac{\tau\tau' - 1}{\tau'(1 + \tau)}.$$

Notice that this value of g is always contained in the domain of $\overline{\Phi}_{\tau\tau'}$. Also, the value of the derivative of $\overline{\Phi}_{\tau\tau'}$ at this fixed point is

$$\left. d\overline{\Phi}_{\tau\tau'} \right|_{g = \frac{\tau\tau'-1}{\tau'(1+\tau)}} = \left. \frac{1}{\tau\tau'\left(1 - g - g/\tau\right)^2} \right|_{g = \frac{\tau\tau'-1}{\tau'(1+\tau)}} = \tau\tau' > 1$$

so we see that $g = \frac{\tau\tau'-1}{\tau'(1+\tau)}$ is always a repelling fixed point of $\overline{\Phi}_{\tau\tau'}$.

To see when this value of g is also a fixed point of $\Phi_{\tau\tau'}$, we need to see for which parameter values it is in the proper domain of definition of $\Phi_{\tau\tau'}$.

Lemma 4.5. *If*

$$\tau' \leq \frac{1 + 2\tau}{\tau^2} \qquad \text{and} \qquad \tau \leq \frac{1 + 2\tau'}{\tau'^2}$$

then

$$g = \frac{\tau\tau' - 1}{\tau'(1 + \tau)}$$

is a fixed point of $\Phi_{\tau\tau'}$.

Proof. First assume that $\tau\tau' - \tau' - 1 \geq 0$. Then we need to solve the inequality

$$\frac{\tau\tau' - 1}{\tau'(1 + \tau)} \leq \frac{\tau}{1 + 2\tau}$$

which has the solution

$$\tau' \leq \frac{1 + 2\tau}{\tau^2}.$$

On the other hand, if we assume that $\tau\tau' - \tau' - 1 \leq 0$, then we need to solve

$$\frac{\tau\tau' - 1}{\tau'(1 + \tau)} \leq \frac{\tau\tau'}{\tau\tau' + 2\tau' + 1}$$

which has the solution

$$\tau \leq \frac{1 + 2\tau'}{\tau'^2}.$$

Now we need to graph the three equations

$$0 = \tau\tau' - \tau' - 1 \tag{1}$$

$$\tau' = \frac{1 + 2\tau}{\tau^2} \tag{2}$$

$$\tau = \frac{1 + 2\tau'}{\tau'^2} \tag{3}$$

so that we can see how all of the above inequalities fit together.

Notice that equation (3) can be rewritten as $\tau\tau'^2 - 2\tau' - 1 = 0$. Then by using the quadratic formula we get

$$\tau' = \frac{2 + \sqrt{4 + 4\tau}}{2\tau} = \frac{1}{\tau} + \sqrt{\frac{1 + \tau}{\tau^2}}$$

which is then easy to compare with equation (2) rewritten in the form

$$\tau' = \frac{1}{\tau} + \frac{1 + \tau}{\tau^2}.$$

So we see that these two equations intersect only when $1 + \tau = \tau^2$, that is, when $\tau = (1 + \sqrt{5})/2$ (and by symmetry, when $\tau' = (1 + \sqrt{5})/2$) and we can also easily tell that when $\tau > (1 + \sqrt{5})/2$ the graph of equation (3) is above the graph of equation (2).

For the equation $0 = \tau\tau' - \tau' - 1$, we can find where it intersects equation (2) by solving

$$\frac{1}{\tau - 1} = \frac{1 + 2\tau}{\tau^2}$$

which is equivalent to $0 = \tau^2 - \tau - 1$ so that they intersect only when $\tau' = (1 + \sqrt{5})/2$.

So the equations (1)–(3) intersect only at the point $((1 + \sqrt{5})/2, (1 + \sqrt{5})/2)$. See Figure 2 for their graphs. From these graphs we see that

$$g = \frac{\tau\tau' - 1}{\tau'(1 + \tau)}$$

is a fixed point for $\Phi_{\tau\tau'}$ only when the inequalities stated in the lemma are true. $\square$

Lemma 4.6. *If*

$$\tau' \leq \frac{1 + 2\tau}{\tau^2} \qquad \text{and} \qquad \tau \leq \frac{1 + 2\tau'}{\tau'^2}$$

then either

$$g = \frac{1}{\tau\tau'} \qquad \text{or} \qquad g = \frac{1}{1 + \tau'}$$

is a fixed point of $\Phi_{\tau\tau'}$ depending on whether the sign of $\tau\tau' - \tau' - 1$ is positive or negative.

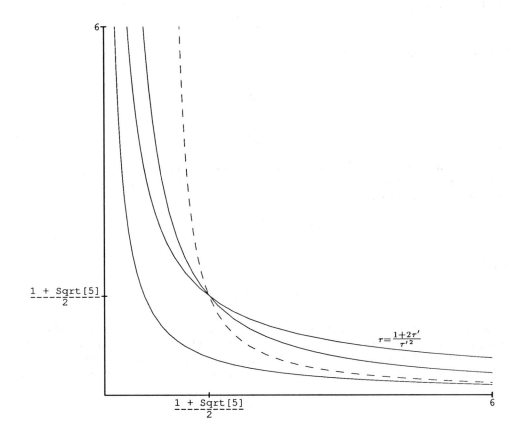

Figure 2

Proof. No matter what sign $\tau\tau' - \tau' - 1$ has, the second part of the definition of $\Phi_{\tau\tau'}$ is a constant function. So we can get an attracting fixed point of $\Phi_{\tau\tau'}$ by solving inequalities so that the value of the constant part lies in the domain of the constant part.

If we assume that $\tau\tau' - \tau' - 1 \geq 0$, then we need to solve the inequality

$$\frac{\tau}{1 + 2\tau} \leq \frac{1}{\tau\tau'}$$

which simplifies to

$$\tau' \leq \frac{1 + 2\tau}{\tau^2}.$$

If we assume that $\tau\tau' - \tau' - 1 \leq 0$, then we need to solve the inequality

$$\frac{\tau\tau'}{\tau\tau' + 2\tau' + 1} \leq \frac{1}{1 + \tau'}$$

which simplifies to

$$\tau \leq \frac{1 + 2\tau'}{\tau'^2}.$$

And so we get the same inequalities as in the previous lemma. $\square$

Lemma 4.7. *If*

$$\tau' > \frac{1 + 2\tau}{\tau^2} \qquad \text{or} \qquad \tau > \frac{1 + 2\tau'}{\tau'^2}$$

then the basin of attraction of zero is all of $[0, 1)$.

Proof. For the function $\overline{\Phi}_{\tau\tau'}$, assume that $g < \frac{\tau\tau' - 1}{\tau'(1+\tau)}$. Then

$$\frac{1}{\tau\tau'(1 - g - g/\tau)} < 1.$$

Let $\epsilon \in \left(\frac{1}{\tau\tau'(1-g-g/\tau)}, 1\right)$. Then

$$\overline{\Phi}_{\tau\tau'}(g) = \frac{g}{\tau\tau'(1 - g - g/\tau)} < \epsilon g.$$

So then $\overline{\Phi}^n_{\tau\tau'}(g) < \epsilon^n g$. So all points g below the fixed point of $\overline{\Phi}_{\tau\tau'}$ are attracted by $\overline{\Phi}_{\tau\tau'}$ to zero at a geometric rate.

Now suppose that $\tau\tau' - \tau' - 1 \geq 0$ and $g \geq \tau/(1 + 2\tau)$. Since zero is the only fixed point, we have

$$\Phi_{\tau\tau'}(g) = \frac{1}{\tau\tau'} < \frac{\tau}{1 + 2\tau} < \frac{\tau\tau' - 1}{\tau'(1 + \tau)}.$$

So $\Phi_{\tau\tau'}(g) < \frac{\tau\tau' - 1}{\tau'(1+\tau)}$ and hence, for some $\epsilon < 1$ we have

$$\Phi^n_{\tau\tau'}(g) = \overline{\Phi}^{n-1}_{\tau\tau'}(\Phi_{\tau\tau'}(g)) < \epsilon^{n-1}\Phi_{\tau\tau'}(g)$$

which tends to zero as $n \to \infty$.

On the other hand, if we suppose that $\tau\tau' - \tau' - 1 \leq 0$ and $g \geq \tau\tau'/(\tau\tau' + 2\tau' + 1)$, since zero is the only fixed point we have

$$\Phi_{\tau\tau'}(g) = \frac{1}{1 + \tau'} < \frac{\tau\tau'}{\tau\tau' + 2\tau' + 1} < \frac{\tau\tau' - 1}{\tau'(1 + \tau)}.$$

So again we have $\Phi^n_{\tau\tau'}(g) < \epsilon^{n-1}\Phi_{\tau\tau'}(g)$. $\square$

Now we now enough about the dynamics of $\Phi_{\tau\tau'}$ to determine when the geometric process from which $\Phi_{\tau\tau'}$ was derived defines Cantor sets and when it fails to define Cantor sets. The way that the function $\Phi_{\tau\tau'}$ gives us information about the geometric process is through the following formula for the lengths of the bridges L_n.

Proposition 4.8. Let $g_n = \Phi^n_{\tau\tau'}(g)$. If $g_n \leq \tau/(1+2\tau)$ for all $n \geq 0$, then for $n \geq 1$ we have

$$|L_n| = \prod_{k=0}^{n-1} \left[1 - g_k \left(\frac{\tau+1}{\tau} \right) \right].$$

Proof. The proof is by induction. When $n = 1$, we have

$$\begin{aligned}
|L_1| &= 1 - g - |U_1| \\
&= 1 - g - g/\tau \\
&= 1 - g \left(\frac{\tau+1}{\tau} \right) \\
&= 1 - g_0 \left(\frac{\tau+1}{\tau} \right).
\end{aligned}$$

Now suppose that $|L_{n-1}| = \prod_{k=0}^{n-2} \left[1 - g_k \left(\frac{\tau+1}{\tau} \right) \right]$. From the picture

we see that $|L_n| = |L_{n-1}| - |U'_{n-1}| - |U_n|$. By the definition of g_{n-1} we have $|U'_{n-1}| = g_{n-1}|L_{n-1}|$. And since $g_{n-1} \leq \tau/(1+2\tau)$, we have $|U_n| = (g_{n-1}/\tau)|L_{n-1}|$. So

$$\begin{aligned}
|L_n| &= |L_{n-1}| - g_{n-1}|L_{n-1}| - \frac{g_{n-1}}{\tau}|L_{n-1}| \\
&= \left(1 - g_{n-1} - \frac{g_{n-1}}{\tau} \right) |L_{n-1}| \\
&= \left(1 - g_{n-1} \left(\frac{\tau+1}{\tau} \right) \right) \prod_{k=0}^{n-2} \left[1 - g_k \left(\frac{\tau+1}{\tau} \right) \right] \\
&= \prod_{k=0}^{n-1} \left[1 - g_k \left(\frac{\tau+1}{\tau} \right) \right]. \qquad\qquad \square
\end{aligned}$$

Proposition 4.9.

(1) If
$$\tau' \leq \frac{1+2\tau}{\tau^2} \qquad \text{and} \qquad \tau \leq \frac{1+2\tau'}{\tau'^2},$$

then the geometric process will define two Cantor sets.

(2) If
$$\tau' > \frac{1+2\tau}{\tau^2} \qquad \text{or} \qquad \tau > \frac{1+2\tau'}{\tau'^2},$$

then the geometric process will always fail to define Cantor sets.

Proof. Assume that the inequalities in item (1) are satisfied. Let $g = \frac{\tau\tau'-1}{\tau'(1+\tau)}$. Then g is a fixed point for $\Phi_{\tau\tau'}$ and $g < \tau/(1+2\tau)$ so $g_n = g$ and $g_n < \tau/(1+2\tau)$ for all n. Obviously, the infinite series

$$\sum_{n=0}^{\infty} g_n$$

diverges, so the infinite product

$$\prod_{n=0}^{\infty} \left(1 - g_n \left(\frac{1+\tau}{\tau}\right)\right)$$

is equal to zero. So then, by the previous proposition, $|L_n| \to 0$ as $n \to \infty$. Also, $|L'_n| < |L_n|$ for all n so $|L'_n| \to 0$ as $n \to \infty$. So the geometric process, when started with the value of $g = \frac{\tau\tau'-1}{\tau'(1+\tau)}$, does define Cantor sets in this case.

Now assume that the inequalities in item (2) are true. Let $g \in (0,1)$ be arbitrary. Since zero has all of $[0,1)$ as its basin of attraction for $\Phi_{\tau\tau'}$, $\Phi_{\tau\tau'}^n(g) \to 0$ as $n \to \infty$. So for n sufficiently large, $g_n < \tau/(1+2\tau)$. So we can assume that $g_0 < \tau/(1+2\tau)$. We have already shown, given the current hypothesis on τ and τ', that $g_n < \epsilon^n g_0$ for some $\epsilon \in (0,1)$. So the infinite series $\sum g_n$ converges. This implies that the infinite product $\prod \left(1 - g_n \left(\frac{1+\tau}{\tau}\right)\right)$ converges to a nonzero number. Then by the previous proposition, $|L_n|$ does not converge to zero. So the geometric process does not define Cantor sets in this case. $\square$

Note. When the inequalities in item (1) are satisfied and when $g_0 = \frac{\tau\tau'-1}{\tau'(1+\tau)}$, the fact that g is a fixed point of $\Phi_{\tau\tau'}$ implies that the geometric process is "self-similar." Each iterate of the process is the same as the previous iterate, just on a different scale. We could also start the geometric process with g_0 equal to the other fixed point of $\Phi_{\tau\tau'}$ and get a different self-similar construction of two interleaved Cantor sets. Since $g = \frac{\tau\tau'-1}{\tau'(1+\tau)}$ is a repelling fixed point of $\Phi_{\tau\tau'}$, if we start the geometric process with $g_0 < \frac{\tau\tau'-1}{\tau'(1+\tau)}$, then the geometric process will fail to define Cantor sets (since $\Phi_{\tau\tau'}^n(g_0)$ will converge to zero at a geometric rate). However, if we start the process with $g_0 > \frac{\tau\tau'-1}{\tau'(1+\tau)}$, then $\Phi_{\tau\tau'}^n(g_0)$ will converge to the other fixed point of $\Phi_{\tau\tau'}$, which is attracting, and the geometric process will not be self-similar but it will define two interleaved Cantor sets whose construction tends towards self-similarity.

When the geometric process does define two Cantor sets, their intersection is the point 0 and a countable number of endpoints of bridges. Using a version of Williams' "shaving lemma", we can get rid of the endpoints, which then proves the second part of Theorem 1.

Proposition 4.10. *If $(\tau, \tau') \in \Lambda_4$, then there exist Cantor sets Γ and Γ' with thicknesses τ and τ' and fundamental intervals $[0,1]$ and $[0, 1-g]$, for some $g \in (0,1)$, such that $\Gamma \cap \Gamma' = \{0\}$.*

Proof. Choose $(\tau, \tau') \in \Lambda_4$ and let $g_0 = \frac{\tau\tau'-1}{\tau'(1+\tau)}$. We will modify the geometric process slightly to "shave off" the endpoints at the expense of the thickness of the Cantor sets.

Choose $\epsilon > 0$. The new geometric process has the following description. Remove gaps U_n and U'_n from bridges L_{n-1} and L'_{n-1} as before. Now move the right hand endpoint of gap U_n just enough to the right so that the new gap, call it V_n, determines thickness $\tau - \epsilon$. Next, move the right hand endpoint of gap U_n' just enough to the right so that the new gap, V_n', determines thickness $\tau' - \epsilon$. See the following picture.

Now the new bridge R_n' lies completely in the new gap V_n so there are no more common endpoints. Also, the bridges L_n and L_n' remain exactly the same as in the old geometric process so the self-similar structure of the process has not been changed and this new process can be iterated as before. $\square$

Now we shall prove the first part of Theorem 1, the part that is a partial statement of the Cantor Lemma. We will actually prove the contrapositive of the statement.

Proposition 4.11. *Suppose that Γ and Γ' are Cantor sets with fundamental intervals $[0,1]$ and $[0, 1-g]$ with $g \in (0,1)$. Let $\mathcal{U}$ and $\mathcal{U}'$ be any orderings of the gaps of Γ and Γ' by decreasing size. Suppose that for all n, $C_n(\mathcal{U})$ and $C_n(\mathcal{U}')$ have only one pair of interleaved bridges. If τ and τ' are the thicknesses of Γ and Γ' and $\tau\tau' > 1$, then $(\tau, \tau') \notin \Lambda_3$.*

Proof. Suppose Γ and Γ' satisfy the hypothesis of the proposition. We will define two subsequences of gaps $\{V_i\}_{i=1}^{\infty}$ and $\{V_i'\}_{i=1}^{\infty}$ for Γ and Γ' respectively that mimic the subsequences of gaps $\{U_i\}_{i=1}^{\infty}$ and $\{U_i'\}_{i=1}^{\infty}$ constructed by the geometric process that begins with $g_0 = g$. By comparing the gaps V_i and V_i' derived from the given Cantor sets with the gaps U_i and U_i' constructed by the geometric process, we will be able to prove the proposition.

Let V_1 be the longest gap of Γ that intersects the interval $[0, 1-g]$ (it may not be true that V_1 is the longest gap of Γ). Let B_1 be the left hand bridge and A_1 the right hand bridge determined by V_1 in the interval $[0,1]$.

Claim. The point $1-g$ is either contained in the gap V_1 or it is the right hand endpoint of V_1.

Suppose not. Then $V_1 \subset (0, 1-g)$. Remove from $[0,1]$ all of the gaps of Γ that are longer than V_1. The point $1-g$ will be in one of the remaining bridges and this bridge will be interleaved with $[0, 1-g]$. But this implies (by Lemmas 2.2 and 2.6) that $C_n(\mathcal{U})$ and $C_n(\mathcal{U}')$ always have at least two pairs of interleaved bridges. This contradicts one of the hypothesis of the proposition, so the claim must be true.

Let V_1' be the longest gap of Γ' that intersects the bridge B_1. Let B_1' be the left hand bridge and A_1' the right hand bridge determined by V_1' in the interval $[0, 1-g]$.

Claim. The left hand endpoint of V_1 is either contained in the gap V_1' or it is the right hand endpoint of V_1'.

The proof is similar to the proof of the previous claim.

Once the gaps V_1 and V_1' are removed from $[0,1]$ and $[0, 1-g]$ we can remove gaps V_2 and V_2' from B_1 and B_1' in an analogous manner by starting over again with B_1 replacing $[0,1]$

and with B_1' replacing $[0, 1 - g]$. In an obvious way, we get two sequences of gaps $\{V_i\}_{i=1}^\infty$ and $\{V_i'\}_{i=1}^\infty$ that are subsequences of all the gaps for Γ and Γ'. By their definitions, the sequences $\{V_i\}_{i=1}^\infty$ and $\{V_i'\}_{i=1}^\infty$ are strictly ordered by size, that is, $|V_{n+1}| < |V_n|$ for all n and similarly for $\{V_i'\}_{i=1}^\infty$. This fact will be important latter in the proof.

Now we need to define a sequence of numbers $\{h_n\}_{n=0}^\infty$, derived from the given Cantor sets Γ and Γ', that mimics the definition of the sequence $g_n = \Phi_{\tau\tau'}^n(g)$ derived from the geometric process.

Let

$$h_0 = g \qquad \text{and} \qquad h_n = \frac{|B_n| - |B_n'|}{|B_n|}.$$

Before continuing with the proof of the proposition, we need to prove two lemmas.

Lemma 4.12. $h_n \leq g_n$ for all n.

Proof. The proof is by induction. Since $h_0 = g = g_0$, the lemma is true for $n = 0$.
Suppose that $h_{n-1} \leq g_{n-1}$.

Claim. $h_n \leq \Phi_{\tau\tau'}(h_{n-1})$.
Before proving the claim, notice that by combining the fact that the function $\Phi_{\tau\tau'}$ is increasing with the induction hypothesis we get $\Phi_{\tau\tau'}(h_{n-1}) \leq \Phi_{\tau\tau'}(g_{n-1}) = g_n$ which, along with the claim, completes the induction step of the proof of the lemma.

Now we shall prove the claim. We start with the following two pictures.

The picture on the left represents removing gaps V_n and V_n' from the intervals B_{n-1} and B_{n-1}'. The picture on the right represents applying the geometric process to the intervals B_{n-1} and B_{n-1}' which gives us new gaps W and W'. By the definition of the geometric process, W determines thickness exactly τ in B_{n-1} and W' determines thickness exactly τ' in B_{n-1}'. Since $\mathcal{V} = \{V_i\}_{i=1}^\infty$ and $\mathcal{V}' = \{V_i'\}_{i=1}^\infty$ are ordered by decreasing size, and since Γ and Γ' have thicknesses τ and τ', Lemma 2.5 tells us that $\tau_n(\mathcal{V}) \geq \tau$ and $\tau_n(\mathcal{V}') \geq \tau'$. In other words, the thickness determined by V_n is greater than or equal to the thickness determined by W and the thickness determined by V_n' is greater than or equal to the thickness determined by W'.

Let L and R be the left and right hand bridges determined by W. Let L' and R' be the left and right hand bridges determined by W'.

Notice that $|B_n| \geq |L|$. (It is easy to see that $|A_n| \leq |R|$. If we had $|B_n| < |L|$, then we would also have $|V_n| > |W|$. But this would contradict that $\tau_n(\mathcal{V}) \geq \tau$.) The fact that $|B_n| \geq |L|$, combined with the fact that the thickness determined by V_n' is greater than or equal to the thickness determined by W', implies that the center of the gap V_n' is more to the right than the center of the gap W'. Using the formulas that are in Section 2 between equation (1) and Lemma 2.9, we can conclude that

$$\frac{|B_n'|}{|V_n'|} \geq \frac{|L'|}{|W'|}.$$

But we also have

$$\frac{|B_n'|}{|B_n| - |B_n'|} \geq \frac{|B_n'|}{|V_n'|}.$$

So we can conclude that

$$\frac{|B_n'|}{|B_n| - |B_n'|} \geq \frac{|L'|}{|W'|}.$$

Then we get the following series of inequalities:

$$|B_n'||W'| \geq |L'|(|B_n| - |B_n'|)$$

$$|B_n'|(|L| - |L'|) \geq |L'|(|B_n| - |B_n'|)$$

$$|B_n'||L| \geq |L'||B_n|$$

$$\frac{|B_n'|}{|B_n|} \geq \frac{|L'|}{|L|}$$

$$\frac{|B_n| - |B_n'|}{|B_n|} \leq \frac{|L| - |L'|}{|L|}$$

$$h_n \leq \Phi_{\tau\tau'}(h_{n-1}).$$

Which completes the proof of the claim and hence, of the lemma also.   □

Lemma 4.13. $|B_n| \geq |L_n|$ for all $n \geq 1$.

Proof. The proof is again by induction.

For $n = 1$, the proof that $|B_1| \geq |L_1|$ is the same as the proof in the last lemma that $|B_n| \geq |L|$.

Assume that $|B_{n-1}| \geq |L_{n-1}|$.

We start with the following two pictures:

The picture on the left represents removing the gaps V_n and V_n' from the bridges B_{n-1} and B_{n-1}'. The picture on the right represents applying the geometric process to the bridges L_{n-1} and L_{n-1}'. From these two picture we see that

$$\frac{|A_n|}{|B_{n-1}|} \leq \frac{|B_{n-1}| - |B_{n-1}'|}{|B_{n-1}|} = h_{n-1} \leq g_{n-1} = \frac{|L_{n-1}| - |L_{n-1}'|}{|L_{n-1}|} = \frac{|R_n|}{|L_{n-1}|}.$$

Dividing all of the lengths in the left picture by $|B_{n-1}|$ and dividing all of the lengths in the right picture by $|L_{n-1}|$ has the effect of "normalizing" each of B_{n-1} and L_{n-1} to be bridges with length one. Then we can see that $|A_n|/|B_{n-1}| \leq |R_n|/|L_{n-1}|$ means that the normalized gap $V_n/|B_{n-1}|$ is centered more to the right in $[0, 1]$ than the normalized gap $U_n/|L_{n-1}|$. Since $V_n/|B_{n-1}|$ determines a thickness at least as large as that determined by $U_n/|L_{n-1}|$, then $V_n/|B_{n-1}|$ being centered more to the right than $U_n/|L_{n-1}|$ implies that

$$\frac{|B_n|}{|B_{n-1}|} \geq \frac{|L_n|}{|L_{n-1}|}.$$

But by the induction hypothesis

$$\frac{|L_n|}{|L_{n-1}|} \geq \frac{|L_n|}{|B_{n-1}|}.$$

So

$$\frac{|B_n|}{|B_{n-1}|} \geq \frac{|L_n|}{|B_{n-1}|}$$

or $|B_n| \geq |L_n|$, which completes the induction proof of the lemma. $\qquad\square$

Now we can continue with the proof by contradiction of the proposition. Suppose that $(\tau, \tau') \in \Lambda_3$. Then from Proposition 9 we know that the lengths of the bridges L_n do not go to zero as $n \to \infty$. Then by the last lemma we know that the lengths of the bridges B_n also do not go to zero as $n \to \infty$. So there must be a $\delta > 0$ such that $[0, \delta] \subset B_n$ for all n. Now we will show that there is no gap of the Cantor set Γ contained in $[0, \delta]$, which is an obvious contradiction to the fact that Γ is a Cantor set.

Suppose there is a gap V of Γ contained in the interval $[0, \delta]$. Since the lengths of the gaps V_n must go to zero as $n \to \infty$, there is an N such that the length of V_N is strictly less than the length of V. So we have the following picture:

But by the definition of V_N, it is the longest gap of Γ that intersects B'_{N-1} which contradicts that $|V_N| < |V|$. So there is no gap of Γ contained in the interval $[0, \delta]$ which contradicts that Γ is a Cantor set. So it must be that $(\tau, \tau') \notin \Lambda_3$. $\qquad\square$

Now we will prove the third and last part of Theorem 1. This result will not be needed anywhere in this paper but the idea behind the proof will be used again.

If (τ, τ') is on the boundary between sets Λ_3 and Λ_4, and if we let $g = \frac{\tau\tau'-1}{\tau'(1+\tau)}$, which is the only fixed point of $\Phi_{\tau\tau'}$ for these parameter values, then the geometric process will construct Cantor sets Γ and Γ' with a countable intersection. We need to prove that for these thicknesses the intersection is always an infinite set.

Proposition 4.14. *Suppose that Γ and Γ' are Cantor sets with thicknesses τ and τ' and with fundamental intervals $[0, 1]$ and $[0, 1-g]$ with $g \in (0, 1)$. If (τ, τ') is on the boundary between Λ_3 and Λ_4, then $\Gamma \cap \Gamma'$ must have infinite cardinality.*

Proof. Suppose that $\Gamma \cap \Gamma'$ has finite cardinality. Obviously $0 \in \Gamma \cap \Gamma'$. Since the intersection set is finite, there is a δ such that if $x \in [0, \delta] \cap (\Gamma \cap \Gamma')$ then $x = 0$. Choose an N large enough such that the components of $C_N(\mathcal{U})$ and $C_N(\mathcal{U}')$ that contain 0 are contained in the interval $[0, \delta]$ (where $\mathcal{U}$ and $\mathcal{U}'$ are any orderings of all the gaps of Γ and Γ' by decreasing size). Let these components be denoted by B and B'. Then $(B \cap \Gamma) \cap (B' \cap \Gamma') = \{0\}$. This means that without lose of generality we can assume that $\Gamma \cap \Gamma' = \{0\}$.

As in the proof of the previous proposition, we can define subsequences of gaps $\{V_i\}_{i=1}^{\infty}$ and $\{V_i'\}_{i=1}^{\infty}$ of Γ and Γ' that are ordered by decreasing size. However, in this case, since $\Gamma \cap \Gamma' = \{0\}$ we can conclude that the gaps V_n and V_n' do not have common endpoints with any bridges. This gives us the stronger inequality $h_n < g_n$ for all $n \geq 1$. So, in particular

$$h_1 < g_1 = \Phi_{\tau\tau'}(g) \leq \frac{\tau\tau' - 1}{\tau'(1 + \tau)}$$

where the last inequality comes from the fact that we have chosen parameter values for $\Phi_{\tau\tau'}$ right at the "saddle-node" bifurcation, so all values of g greater than $\frac{\tau\tau'-1}{\tau'(1+\tau)}$ are mapped to $\frac{\tau\tau'-1}{\tau'(1+\tau)}$ and all values less than $\frac{\tau\tau'-1}{\tau'(1+\tau)}$ are mapped strictly closer to zero. So, without lose of generality, we can assume that $g < \frac{\tau\tau'-1}{\tau'(1+\tau)}$.

But now we know that the geometric process fails, because g is in the basin of attraction of the hyperbolic sink at the origin. The lengths of the L_n do not go to zero and therefore the lengths of the B_n do not go to zero either. And, as in the proof of the last proposition, this contradicts that Γ is a Cantor set. So it must be that $\Gamma \cap \Gamma'$ has infinite cardinality. $\square$

5. Second Kind Of Overlapped Point

In this section we will consider the possibility of two Cantor sets Γ and Γ' with $\Gamma \cap \Gamma' = \{x\}$ where x is an overlapped point of the second kind.

Let Γ and Γ' be Cantor sets with fundamental intervals $[-1, 1]$ and $[0, 1 - g]$ with $g \in (0, 1)$. Suppose that $\Gamma \cap \Gamma' = \{0\}$ where 0 is an overlapped point of the second kind. Then $\Gamma \cap [0, 1]$ and Γ' make up a pair of Cantor sets like those studied in the last section. Let $\hat{\tau}$ be the thickness of $\Gamma \cap [0, 1]$ and let τ' be the thickness of Γ'. Then from Theorem 1 of the previous section we know that $(\hat{\tau}, \tau') \in \Lambda_4$. Let τ be the thickness of Γ. It is possible that $\tau \geq \hat{\tau}$, since if U is a gap of Γ that lies in $[0, 1]$, then U will have a longer left hand bridge in Γ then it will have in $\Gamma \cap [0, 1]$. The following theorem describes how much greater τ can be than $\hat{\tau}$ (see also the statement of Theorem 6 at the end of this section).

Theorem 5.1. *Define sets Λ_5 and Λ_6 by*

$$\Lambda_5 = \left\{ (\tau, \tau') \,\Big|\, \tau > \frac{1 + 2\tau'}{\tau'^2} \right\}$$

$$\Lambda_6 = \{ (\tau, \tau') \mid \tau\tau' > 1 \} \setminus \text{closure}(\Lambda_5).$$

Let Γ and Γ' be Cantor sets with fundamental intervals $I = [-1, 1]$ and $I' = [0, 1-g]$ with $g \in (0, 1)$. Let τ and τ' be their thicknesses. Suppose that $0 \in \Gamma \cap \Gamma'$ and that 0 is an overlapped point of the second kind.

(1) *If $(\tau, \tau') \in \Lambda_5$, then $\Gamma \cap \Gamma' \neq \{0\}$. In fact there is an n such that C_n and C_n' have two pairs of interleaved bridges.*
(2) *If $(\tau, \tau') \in \Lambda_6$, then it may be that $\Gamma \cap \Gamma' = \{0\}$.*

We will prove this theorem by defining a new geometric process. We will start with the intervals $I = [-1, 1]$ and $I' = [0, 1-g]$ with $g \in (0, 1)$, and with two positive numbers τ and τ' with $\tau\tau' > 1$. There are two cases to be considered when defining this new geometric process, when $g < \tau/(1 + \tau)$ and when $g \geq \tau/(1 + \tau)$. First we will assume the case that $g < \tau/(1 + \tau)$.

The first step of the first iterate of the geometric process in this case is to remove a gap U_1 from I such that the right hand endpoint of U_1 is the point $1 - g$ and so that U_1 determines thickness τ in the interval $[-1, 1]$ (the condition $g < \tau/(1 + \tau)$ implies that $g + |U_1| < 1$, so $0 \notin U_1$). The second step of the first iterate is to remove a gap U_1' from the interval I' so that the right hand endpoint of U_1' is equal to the left hand endpoint of U_1 and so that U_1' determines thickness τ' in I'. The third step of the first iterate is to remove a gap U_1'' from the interval $[-1, 0]$ so that the distance from the right hand endpoint of U_1'' to 0 is the same as the distance from 0 to the left hand endpoint of U_1 and so that U_1''

49

determines thickness τ in the interval from -1 to the left hand endpoint of U_1. At this point we have the following picture.

$$
\begin{array}{cccccc}
-1 & U_1'' & 0 & U_1 & R_1 & 1 \\
\vdash\!\!\!\dashv & \vdash\!\!\!\dashv\!\!\!\rule{2cm}{0.4pt}\!\!\!\dashv & & \vdash\!\!\!\dashv\!\!\!\dashv & & \\
& & & \vdash\!\!\!\rule{2cm}{0.4pt}\!\!\!\dashv & \vdash\!\!\!\dashv\!\!\!\dashv & \\
& & & U_1' & R_1' & 1-g
\end{array}
$$

The fourth and last step of the first iterate is to construct a middle-α Cantor set with thickness τ in the bridge R_1, a middle-α Cantor set with thickness τ' in the bridge R_1' and a middle-α Cantor set with thickness τ in the bridge to the left of U_1''.

Notice that in the interval $[0, 1]$ this geometric process is very similar to the geometric process from the previous section. It attempts to construct two Cantor sets in $[0, 1]$ with 0 as the only overlapped point. The difference between the restriction of this new geometric process to $[0, 1]$ and the geometric process from the previous section is that this new process, in the interval $[0, 1]$, never has to consider the left hand bridge of the gap U_1. The right hand bridge of U_1 is always the one that is used to determine the thickness.

Now we will translate this case of the new geometric process into a dynamical system. Notice that we always have

$$
|U_1| = \frac{|R_1|}{\tau} = \frac{g}{\tau}.
$$

(This is where the condition on g comes from so that $0 \notin U_1$ in this case.) The equation for $|U_1'|$ is given by

$$
|U_1'| = \begin{cases} \dfrac{|R_1'|}{\tau'} & \text{if } |R_1'| \le \dfrac{\tau'}{1+2\tau'}(1-g) \\[2ex] \dfrac{|L_1'|}{\tau'} & \text{if } |R_1'| \ge \dfrac{\tau'}{1+2\tau'}(1-g) \end{cases}
$$

$$
= \begin{cases} \dfrac{|U_1|}{\tau'} & \text{if } |U_1| \le \dfrac{\tau'}{1+2\tau'}(1-g) \\[2ex] \dfrac{1-g-|U_1|-|U_1'|}{\tau'} & \text{if } |U_1| \ge \dfrac{\tau'}{1+2\tau'}(1-g). \end{cases}
$$

By solving for $|U_1'|$ in the second part of the equation we get

$$
|U_1'| = \begin{cases} \dfrac{|U_1|}{\tau'} & \text{if } |U_1| \le \dfrac{\tau'}{1+2\tau'}(1-g) \\[2ex] \dfrac{1-g-|U_1|}{1+\tau'} & \text{if } |U_1| \ge \dfrac{\tau'}{1+2\tau'}(1-g). \end{cases}
$$

Then letting $|U_1| = g/\tau$ we get

$$
|U_1'| = \begin{cases} \dfrac{g}{\tau\tau'} & \text{if } g \le \dfrac{\tau\tau'}{\tau\tau'+2\tau'+1} \\[2ex] \dfrac{1-g-g/\tau}{1+\tau'} & \text{if } g \ge \dfrac{\tau\tau'}{\tau\tau'+2\tau'+1}. \end{cases}
$$

After having removed the gaps U_1, U_1' and U_1'', let L_1 denote the interval from 0 to the left hand endpoint of U_1. Then we rescale everything by dividing by $|L_1|$. This gives us a new value for g that we will denote $\hat{\Phi}_{\tau\tau'}(g)$. The formula for $\hat{\Phi}_{\tau\tau'}$ is given by

$$\hat{\Phi}_{\tau\tau'}(g) = \frac{|U_1'|}{1 - g - |U_1|}$$

$$= \begin{cases} \dfrac{g}{\tau\tau'(1 - g - g/\tau)} & \text{if } g \le \dfrac{\tau\tau'}{\tau\tau' + 2\tau' + 1} \\[2ex] \dfrac{1}{1 + \tau'} & \text{if } \dfrac{\tau\tau'}{\tau\tau' + 2\tau' + 1} \le g < \dfrac{\tau}{1 + \tau}. \end{cases}$$

So far we have defined $\hat{\Phi}_{\tau\tau'}$ so that its domain is $g \in [0, \tau/(1 + \tau))$. Notice that we always have $\tau\tau'/(\tau\tau' + 2\tau' + 1) < \tau/(1 + \tau)$ and that $1/(1 + \tau') < \tau/(1 + \tau)$ since $\tau\tau' > 1$.

Now we will consider the case when $g \ge \tau/(1 + \tau)$. If we try to define the first step of the first iterate of the geometric process in this case the same way as in the last case, then we will have $0 \in U_1$. But we want 0 to be in the intersection of the two Cantor sets. So choose an $\epsilon > 0$ (how to choose ϵ is described in the next step) and let U_1 be the open interval $(\epsilon, 1 - g)$. Notice that this makes the thickness determined by U_1 in $[-1, 1]$ strictly greater than τ. The second step of the first iterate is to remove a gap U_1' from the interval I' so that the right hand endpoint of U_1' is equal to ϵ and so that U_1' determines thickness τ' in $[0, 1 - g]$. But we also want to specify that the left hand bridge of U_1' in $[0, 1 - g]$ should be the bridge that determines the thickness τ'. So ϵ should be chosen so that

$$\epsilon < \frac{1 + \tau'}{1 + 2\tau'}(1 - g).$$

The third step of the first iterate is to remove a gap U_1'' from the interval $[-1, \epsilon]$ so that its right hand endpoint is equal to $-\epsilon$ and such that U_1'' determines thickness τ in $[-1, \epsilon]$. Finally, the fourth step of the first iterate is to define a middle-α Cantor set with thickness τ in the right hand bridge of U_1, a middle-α Cantor set with thickness τ' in the right hand bridge of U_1' and a middle-α Cantor set with thickness τ in the left hand bridge of U_1''.

Now we will translate this case of the geometric process into a function (that is, we will extend the definition of $\hat{\Phi}_{\tau\tau'}$ to the interval $[\tau/(1 + \tau), 1)$).

Let a denote the left hand endpoint of U_1'.

Then $|U_1'| = \epsilon - a$ and

$$\hat{\Phi}_{\tau\tau'}(g) = \frac{|U_1'|}{\epsilon} = \frac{\epsilon - a}{\epsilon}.$$

We need to compute the value the of a. Suppose that the value of ϵ is chosen sufficiently small so that the left hand bridge of U_1' determines the thickness τ'. Then

$$\tau' = \frac{a}{\epsilon - a}$$

so

$$a = \frac{\epsilon \tau'}{1 + \tau'}.$$

Then

$$|U_1'| = \epsilon - a = \frac{\epsilon}{1 + \tau'}$$

and finally,

$$\hat{\Phi}_{\tau\tau'}(g) = \frac{|U_1'|}{\epsilon} = \frac{1}{1 + \tau'}.$$

So the extension of the definition of $\hat{\Phi}_{\tau\tau'}$ to the interval $[\tau/(1+\tau), 1)$ is just the constant function equal to $1/(1 + \tau')$. The definition of $\hat{\Phi}_{\tau\tau'}$ on the domain $[0, 1)$ is then given by

$$\hat{\Phi}_{\tau\tau'}(g) = \begin{cases} \dfrac{g}{\tau\tau'(1 - g - g/\tau)} & \text{if } g \leq \dfrac{\tau\tau'}{\tau\tau' + 2\tau' + 1} \\[3mm] \dfrac{1}{1 + \tau'} & \text{if } g \leq \dfrac{\tau\tau'}{\tau\tau' + 2\tau' + 1}. \end{cases}$$

Note. Since $\tau\tau' > 1$ implies that $1/(1 + \tau') < \tau/(1 + \tau)$, the second case of the geometric process can only come up at the first iterate. No matter what the initial choice of g is, after one iteration of the geometric process we only need to worry about the first case.

The dynamics of $\hat{\Phi}_{\tau\tau'}$ are described by the following proposition.

Lemma 5.2. *Assume that $\tau\tau' > 1$.*

(1) *Zero is a hyperbolic sink for $\hat{\Phi}_{\tau\tau'}$.*

(2) *If τ and τ' also satisfy the inequality*

$$\tau \leq \frac{1 + 2\tau'}{\tau'^2},$$

then the points

$$g = \frac{\tau\tau' - 1}{\tau'(1 + \tau)} \quad \text{and} \quad g = \frac{1}{1 + \tau'}$$

are fixed points of $\hat{\Phi}_{\tau\tau'}$.

(3) *If τ and τ' do not satisfy the inequality in the last item, then zero is the only fixed point of $\hat{\Phi}_{\tau\tau'}$ and its basin of attraction is all of $[0, 1)$.*

The proof of this lemma is similar to the proof of Proposition 4.3 (but shorter since the definition of $\hat{\Phi}_{\tau\tau'}$ is simpler than the definition of $\Phi_{\tau\tau'}$).

The way that the function $\hat{\Phi}_{\tau\tau'}$ gives us information about the geometric process is through the following formula for the lengths of the intervals L_n (where here, L_n is the interval from 0 to the left hand endpoint of U_n).

Lemma 5.3. *Let $g_n = \hat{\Phi}^n_{\tau\tau'}(g)$. Then for $n \geq 1$ we have*

$$|L_n| = \begin{cases} \displaystyle\prod_{k=0}^{n-1}\left[1 - g_k\left(\frac{\tau+1}{\tau}\right)\right] & \text{if } g < \dfrac{\tau}{1+\tau} \\[4ex] \displaystyle\epsilon\prod_{k=1}^{n-1}\left[1 - g_k\left(\frac{\tau+1}{\tau}\right)\right] & \text{if } g \geq \dfrac{\tau}{1+\tau}. \end{cases}$$

The proof is the same as the proof of Proposition 4.8 except that here we do not need to put any restrictions on the value of g_n.

Lemma 5.4.

(1) *If*
$$\tau \leq \frac{1+2\tau'}{\tau'^2},$$
then the geometric process will define two Cantor sets.

(2) *If*
$$\tau > \frac{1+2\tau'}{\tau'^2},$$
then the geometric process will always fail to define Cantor sets.

The proof is similar to the proof of Proposition 4.9.

When the geometric process does define two Cantor sets, their intersection is the point 0 and a countable number of endpoints of bridges. These endpoints can be shaved off as in the proof of Proposition 4.10, and this proves the second part of Theorem 1 from this section.

Now we can look at the proof of the first part of Theorem 1.

Proposition 5.5. *Let Γ and Γ' be Cantor sets with fundamental intervals $[-1,1]$ and $[0, 1-g]$ with $g \in (0,1)$. Suppose that 0 is an overlapped point of the second kind. Let $\mathcal{U}$ and $\mathcal{U}'$ be any orderings of the gaps of Γ and Γ' by decreasing size. Suppose that for all n, $C_n(\mathcal{U})$ and $C_n(\mathcal{U}')$ have only one pair of interleaved bridges. If τ and τ' are the thicknesses of Γ and Γ', then $(\tau, \tau') \notin \Lambda_5$.*

The proof is similar to the proof of Proposition 4.11. First derive from the Cantor sets Γ and Γ' the two subsequences of gaps $\{V_n\}_{n=1}^{\infty}$ and $\{V_n'\}_{n=1}^{\infty}$ that mimic the subsequences of gaps $\{U_n\}_{n=1}^{\infty}$ and $\{U_n'\}_{n=1}^{\infty}$ constructed by the geometric process. Define the intervals B_n between the point 0 and the left hand endpoints of the gaps V_n. Define the sequence of numbers $\{h_n\}_{n=0}^{\infty}$ that mimic the numbers $g_n = \hat{\Phi}^n_{\tau\tau'}(g)$. Then prove that $h_n \leq g_n$. The proof is exactly the same as the proof of Lemma 4.12 (the difference between the two geometric processes doesn't effect the proof of this step). Next prove that $|B_n| \geq |L_n|$ for all n. The proof is exactly the same as the proof of Lemma 4.13. Finally, the assumption that $(\tau, \tau') \in \Lambda_5$ implies (because of Lemma 4 above) that the lengths of the intervals L_n from the geometric process do not go to zero which implies that the lengths of the intervals B_n from Γ do not go to zero which contradicts that Γ is a Cantor set.

The last thing we will do in this section is to notice that the sets Λ_5 and Λ_6 are not symmetric with respect to the line $\tau = \tau'$. This is because we arbitrarily added an extra piece of interval to $[0,1]$ instead of to $[0, 1 - g]$. If we allow the extra interval to be put on $[0, 1 - g]$ instead of on $[0,1]$, then we can get the following symmetrical result.

Theorem 5.6. *Define sets Λ_7 and Λ_8 by*

$$\Lambda_7 = \left\{ (\tau, \tau') \,\middle|\, \tau > \frac{1 + 2\tau'}{\tau'^2} \text{ and } \tau' > \frac{1 + 2\tau}{\tau^2} \right\}$$

$$\Lambda_8 = \{ (\tau, \tau') \mid \tau\tau' > 1 \} \setminus \text{closure}(\Lambda_7).$$

Let Γ and Γ' be Cantor sets with thicknesses τ and τ'. Suppose that $x \in \Gamma \cap \Gamma'$ and that x is an overlapped point of the second kind.

    (1) *If $(\tau, \tau') \in \Lambda_7$, then $\Gamma \cap \Gamma' \neq \{x\}$. In fact, there is an n such that C_n and C'_n have two pairs of interleaved bridges.*

    (2) *If $(\tau, \tau') \in \Lambda_8$, then it may be that $\Gamma \cap \Gamma' = \{x\}$.*

Notice that the set Λ_8 contains the set Λ_4 from the previous section. This is because the set Λ_7 has an "and" between the inequalities in its definition and Λ_3 has an "or" between the same inequalities in its definition (see also Figure 2 from the previous section). So the geometric process from this section works for more pairs of thicknesses than does the geometric process from the last section.

6. First Kind Of Overlapped Point

In this section we will begin the analysis of the possibility of two Cantor sets Γ and Γ' with $\Gamma \cap \Gamma' = \{x\}$ where x is an overlapped point of the first kind. This is the most important of the three kinds of overlapped points and it is this case that will give us the main result stated in Theorem 1.1. The analysis of this case will take up the next several sections (up to Section 10). The main result about this case is given by the following theorem.

Theorem 6.1. *Define sets Λ_1 and Λ_2 by*

$$\Lambda_1 = \left\{ (\tau, \tau') \;\middle|\; \tau > \frac{\tau'^2 + 3\tau' + 1}{\tau'^2} \text{ or } \tau' > \frac{\tau^2 + 3\tau + 1}{\tau^2} \right\}$$

$$\cap \left\{ (\tau, \tau') \;\middle|\; \tau > \frac{(1 + 2\tau')^2}{\tau'^3} \text{ or } \tau' > \frac{(1 + 2\tau)^2}{\tau^3} \right\}$$

$$\Lambda_2 = \{ (\tau, \tau') \mid \tau\tau' > 1 \} \setminus \text{closure}(\Lambda_1).$$

Let Γ and Γ' be Cantor sets with fundamental intervals $I = [0, 1 - g]$ and $I' = [g', 1]$ where $g, g' \in (0, 1)$ and $g + g' < 1$. Let τ and τ' be the thicknesses of Γ and Γ'. Suppose that there is an $x \in \Gamma \cap \Gamma'$ such that x is an overlapped point of the first kind.

(1) *If $(\tau, \tau') \in \Lambda_1$, then $\Gamma \cap \Gamma' \neq \{0\}$. In fact, there is an n such that C_n and C'_n will have two pairs of interleaved bridges.*

(2) *If $(\tau, \tau') \in \Lambda_2$, then it may be that $\Gamma \cap \Gamma' = \{x\}$.*

(3) *If (τ, τ') is on the piece of the boundary between Λ_1 and Λ_2 that is defined by the equation*

$$\tau' = \frac{(1 + 2\tau)^2}{\tau^3}$$

with $\tau \in (1 + \sqrt{2}, \tau_0]$, then it may be that $\Gamma \cap \Gamma' = \{x\}$.

(4) *If $(\tau, \tau') = (1 + \sqrt{2}, 1 + \sqrt{2})$ or if (τ, τ') is on the piece of the boundary between Λ_1 and Λ_2 that is defined by the equation*

$$\tau = \frac{\tau'^2 + 3\tau' + 1}{\tau'^2}$$

with $\tau \in (\tau_0, \infty)$, then $\Gamma \cap \Gamma'$ must contain an infinite number of points. However, it may be that $\Gamma \cap \Gamma'$ is a countable set.

We will prove this theorem in a way similar to the main theorem from Section 4 (the section about overlapped points of the third kind). We will define an iterated geometric process that will attempt to construct Cantor sets Γ and Γ', with thicknesses τ and τ', in the fundamental intervals $[0, 1-g]$ and $[g', 1]$. We will translate this geometric process into a function $\Psi_{\tau\tau'}$ that can be iterated. By analyzing the resulting dynamical system we can determine for which thicknesses the geometric process succeeds and for which it fails.

The rest of this section will consist of the definition of the geometric process, the derivation of the map $\Psi_{\tau\tau'}$ and some of the geometric properties of $\Psi_{\tau\tau'}$. The next section will analyze the dynamics of $\Psi_{\tau\tau'}$. The section after that (Section 8) will use $\Psi_{\tau\tau'}$ to analyze the geometric process.

To define the geometric process, we start with two numbers τ and τ' with $\tau\tau' > 1$ and with the two intervals $I = [0, 1-g]$ and $I' = [g', 1]$ where $g, g' \in (0,1)$ and $g + g' < 1$. The first step of the first iterate of the geometric process is to remove a gap U_1 from the interval I such that the left hand endpoint of U_1 is the point g' and so that U_1 determines thickness τ in I. The second step of the first iterate is to remove a gap U_1' from the interval I' such that the right hand endpoint of U_1' is the point $1-g$ and so that U_1' determines thickness τ' in I'. Notice that the left hand endpoint of U_1' is strictly to the right of the right hand endpoint of U_1, otherwise we would have a contradiction of $\tau\tau' > 1$.

The third and last step of the first iterate is to construct a middle-α Cantor set with thickness τ in the left hand bridge of U_1 and to construct a middle-α Cantor set with thickness τ' in the right hand bridge of U_1'.

The second iterate of the geometric process consists of removing a gap U_2 from the bridge R_1 so that the right hand endpoint of U_2 is equal to the left hand endpoint of U_1' and so that U_2 determines thickness τ in R_1 and then removing a gap U_2' from the bridge L_1' with the left hand endpoint of U_2' equal to the right hand endpoint of U_1 and so that U_2' determines thickness τ' in L_1'.

Then construct middle-α Cantor sets in the bridges R_2 and L_2' with thicknesses τ and τ' respectively.

After the second iterate of the geometric process, the bridges L_2 and R_2' are positioned, relative to each other, the same way that the intervals I and I' were positioned. So the process can be started again with L_2 and R_2' as the fundamental intervals.

In the n-th iterate of the geometric process, after the gaps U_n and U_n' have been removed, let B_n and B_n' denote the bridges of U_n and U_n' that are overlapping. So when n is odd, $B_n = R_n$ and $B_n' = L_n'$ and when n is even, $B_n = L_n$ and $B_n' = R_n'$.

The above process will define two Cantor sets if the lengths of the bridges B_n and B'_n go to zero as $n \to \infty$. As in the previous case of overlapped points of the third kind, if the thicknesses τ and τ' are sufficiently large and if g and g' are sufficiently small, then we would expect the lengths of the gaps U_n and U'_n to go to zero so fast that the lengths of the bridges B_n and B'_n will remain bounded away from zero and so the process would fail. When the process does succeed, the two Cantor sets will have as their intersection a countable number of endpoints of bridges plus a point that is an overlapped point of the first kind and which is the limit point of the endpoints. Later we will derive a formula for the value of this point as a function of τ, τ', g and g'. As in Section 4, we will be able to "shave off" the endpoints at the expense of the thicknesses.

In order to analyze this geometric process, we will now translate it into a function $\Psi_{\tau\tau'}$. First we need formulas for the lengths of the gaps U_1 and U'_1.

$$|U_1| = \begin{cases} \dfrac{g'}{\tau} & \text{if } g' \leq (1-g)\dfrac{\tau}{1+2\tau} \\[3mm] \dfrac{1-(g+g')-|U_1|}{\tau} & \text{if } g' \geq (1-g)\dfrac{\tau}{1+2\tau} \end{cases}$$

so

$$|U_1| = \begin{cases} \dfrac{g'}{\tau} & \text{if } g' \leq (1-g)\dfrac{\tau}{1+2\tau} \\[3mm] \dfrac{1-(g+g')}{1+\tau} & \text{if } g' \geq (1-g)\dfrac{\tau}{1+2\tau}. \end{cases}$$

And

$$|U'_1| = \begin{cases} \dfrac{g}{\tau'} & \text{if } g \leq (1-g')\dfrac{\tau'}{1+2\tau'} \\[3mm] \dfrac{1-(g+g')-|U'_1|}{\tau'} & \text{if } g \geq (1-g')\dfrac{\tau'}{1+2\tau'} \end{cases}$$

so

$$|U'_1| = \begin{cases} \dfrac{g}{\tau'} & \text{if } g \leq (1-g')\dfrac{\tau'}{1+2\tau'} \\[3mm] \dfrac{1-(g+g')}{1+\tau'} & \text{if } g \geq (1-g')\dfrac{\tau'}{1+2\tau'}. \end{cases}$$

The inequalities in these formulas are derived using Lemma 2.9.

After removing the gaps U_1 and U'_1, the overlapping intervals R_1 and L'_1 look like where the length of $R_1 \cup L_1'$ is given by $1 - g - g'$. In order to get the above picture into a form that mimics the original relationships between I and I', first flip over the intervals to get

then divide all the lengths by $1 - g - g'$ to rescale the picture so that the total length is now 1. Then we have the picture

Define

$$\Psi_{\tau\tau'}(g,g') = \left(\frac{|U_1|}{1 - g - g'} , \frac{|U_1'|}{1 - g - g'} \right).$$

By plugging in the formulas for $|U_1|$ and $|U_1'|$ we get

$$\Psi_{\tau\tau'}(g,g') = \begin{cases} \left(\frac{g'}{\tau(1-g-g')} , \frac{g}{\tau'(1-g-g')} \right) & \text{if } g' \leq \frac{(1-g)\tau}{1+2\tau} \text{ and } g \leq \frac{(1-g')\tau'}{1+2\tau'} \\[2mm] \left(\frac{g'}{\tau(1-g-g')} , \frac{1-(g+g')}{(1+\tau')(1-g-g')} \right) & \text{if } g' \leq \frac{(1-g)\tau}{1+2\tau} \text{ and } g \geq \frac{(1-g')\tau'}{1+2\tau'} \\[2mm] \left(\frac{1-(g+g')}{(1+\tau)(1-g-g')} , \frac{g}{\tau'(1-g-g')} \right) & \text{if } g' \geq \frac{(1-g)\tau}{1+2\tau} \text{ and } g \leq \frac{(1-g')\tau'}{1+2\tau'} \\[2mm] \left(\frac{1-(g+g')}{(1+\tau)(1-g-g')} , \frac{1-(g+g')}{(1+\tau')(1-g-g')} \right) & \text{if } g' \geq \frac{(1-g)\tau}{1+2\tau} \text{ and } g \geq \frac{(1-g')\tau'}{1+2\tau'}. \end{cases}$$

After some simplification, we get the following formula for the function $\Psi_{\tau\tau'}$.

$$\Psi_{\tau\tau'}(g,g') = \begin{cases} \left(\frac{g'}{\tau(1-g-g')} , \frac{g}{\tau'(1-g-g')} \right) & \text{if } g' \leq \frac{(1-g)\tau}{1+2\tau} \text{ and } g \leq \frac{(1-g')\tau'}{1+2\tau'} \\[3mm] \left(\frac{g'}{\tau(1-g-g')} , \frac{1}{1+\tau'} \right) & \text{if } g' \leq \frac{(1-g)\tau}{1+2\tau} \text{ and } g \geq \frac{(1-g')\tau'}{1+2\tau'} \\[3mm] \left(\frac{1}{1+\tau} , \frac{g}{\tau'(1-g-g')} \right) & \text{if } g' \geq \frac{(1-g)\tau}{1+2\tau} \text{ and } g \leq \frac{(1-g')\tau'}{1+2\tau'} \\[3mm] \left(\frac{1}{1+\tau} , \frac{1}{1+\tau'} \right) & \text{if } g' \geq \frac{(1-g)\tau}{1+2\tau} \text{ and } g \geq \frac{(1-g')\tau'}{1+2\tau'}. \end{cases}$$

From now on the domain of $\Psi_{\tau\tau'}$ will be

$$\{ (g,g') \mid g, g' \geq 0, \; g + g' < 1 \}$$

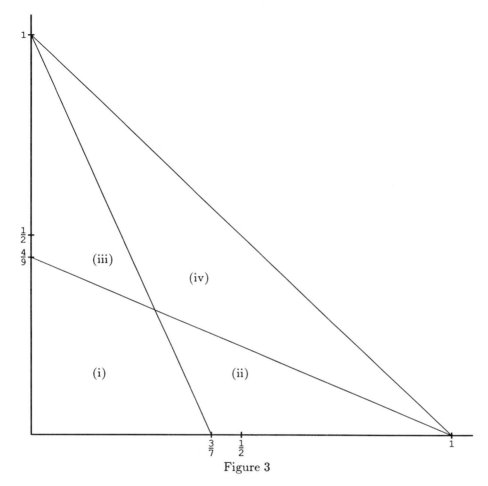

Figure 3

even though the lines $g = 0$ and $g' = 0$ have no meaning for the geometric process from which $\Psi_{\tau\tau'}$ was defined. We will denote the four domains of definition for $\Psi_{\tau\tau'}$ by (i), (ii), (iii), and (iv). See Figure 3 for a picture of the domain of $\Psi_{\tau\tau'}$ (when $\tau = 4$ and $\tau' = 3$).

Notice that $\Psi_{\tau\tau'}$ is a continuous function but it is only piecewise differentiable. Now we shall give some lemmas that describe the geometry of the function $\Psi_{\tau\tau'}$.

Lemma 6.2. *The first coordinate of $\Psi_{\tau\tau'}$ is constant along lines through the point $(1,0)$.*

Proof. In regions (iii) and (iv) the first coordinate of $\Psi_{\tau\tau'}$ is a constant equal to $1/(1+\tau)$.

In regions (i) and (ii), the first coordinate of $\Psi_{\tau\tau'}$ will be constant if

$$\frac{g'}{1 - g - g'} = k$$

for some constant k. Solving for g' as a function of g, we get that the first coordinate of

$\Psi_{\tau\tau'}$ in regions (i) and (ii) is constant on the line

$$g' = \left(\frac{k}{1+k}\right)(1-g)$$

for any constant k. The slope of this line will vary from 0 to $-\tau/(1+2\tau)$ as k varies from 0 to $\tau/(1+\tau)$. The point $(1,0)$ is clearly always on this line. $\square$

Similarly, we have the following lemma.

Lemma 6.3. *The second coordinate of $\Psi_{\tau\tau'}$ is constant along lines through the point $(0,1)$.*

Corollary 6.4. *In region (ii), the function $\Psi_{\tau\tau'}$ is constant along lines through the point $(1,0)$. In region (iii), $\Psi_{\tau\tau'}$ is constant along lines through the point $(0,1)$.*

Lemma 6.5. *If we foliate the domain of $\Psi_{\tau\tau'}$ with lines through the point $(1,0)$, then the image under $\Psi_{\tau\tau'}$ of this foliation will be a foliation of the image of $\Psi_{\tau\tau'}$ by vertical lines.*

Lemma 6.6. *If we foliate the domain of $\Psi_{\tau\tau'}$ with lines through the point $(0,1)$, then the image under $\Psi_{\tau\tau'}$ of this foliation will be a foliation of the image of $\Psi_{\tau\tau'}$ by horizontal lines.*

Lemma 6.7. *The image of $\Psi_{\tau\tau'}$ is equal to the image of region (i).*

Proof. Since $\Psi_{\tau\tau'}$ in region (iii) is constant along lines through the point $(0,1)$, the image of region (iii) is the same as the image of the top edge of region (i).

Since $\Psi_{\tau\tau'}$ is constant in region (iv), the image of region (iv) is the same as the image of the upper right hand corner of region (i).

Since $\Psi_{\tau\tau'}$ in region (ii) is constant along lines through the point $(1,0)$, the image of region (ii) is the same as the image of the right hand edge of region (i). $\square$

Lemma 6.8. *The image under $\Psi_{\tau\tau'}$ of region (i) is a rectangle that has its lower left hand corner at the origin. The upper edge of the rectangle is the image of the right hand edge of region (i). The right hand edge of the rectangle is the image of the top edge of region (i).*

Proof. The proof is an easy consequence of the last three lemmas. See Figure 4 for a picture of the image of $\Psi_{\tau\tau'}$ (when again, $\tau = 4$ and $\tau' = 3$). $\square$

We will say that $(g_1, g_1') \leq (g_2, g_2')$ iff $g_1 \leq g_2$ and $g_1' \leq g_2'$. The next lemma shows that $\Psi_{\tau\tau'}$ preserves this order relation on its domain.

Lemma 6.9. *If $(g_1, g_1') \leq (g_2, g_2')$, then $\Psi_{\tau\tau'}(g_1, g_1') \leq \Psi_{\tau\tau'}(g_2, g_2')$.*

Proof. Let π_1 and π_2 denote the projection maps on the plane. So $\pi_1 \circ \Psi_{\tau\tau'}$ and $\pi_2 \circ \Psi_{\tau\tau'}$ are the first and second components of $\Psi_{\tau\tau'}$.

Suppose that $(g_1, g_1') \leq (g_2, g_2')$. We will show that

$$\pi_1(\Psi_{\tau\tau'}(g_1, g_1')) \leq \pi_1(\Psi_{\tau\tau'}(g_2, g_2'))$$

(the proof for the second coordinate of $\Psi_{\tau\tau'}$ is similar). Let l_1 and l_2 be the lines through the point $(1,0)$ and the points (g_1, g_1') and (g_2, g_2'). By the hypothesis $(g_1, g_1') \leq (g_2, g_2')$,

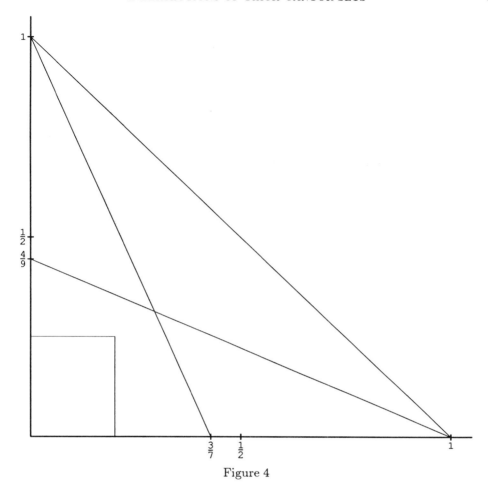

Figure 4

we know that the absolute value of the slope of l_1 is less than or equal to the absolute value of the slope of l_2. As a function of $|m|$, the absolute value of the slope of a line through the point $(1,0)$, the first coordinate of $\Psi_{\tau\tau'}$ has the following equation

$$\pi_1(\Psi_{\tau\tau'}(|m|)) = \begin{cases} \dfrac{|m|}{\tau(1-|m|)} & \text{if } |m| \leq \dfrac{\tau}{1+2\tau} \\[2ex] \dfrac{1}{1+\tau} & \text{if } |m| \geq \dfrac{\tau}{1+2\tau}. \end{cases}$$

So we see that the first coordinate of $\Psi_{\tau\tau'}$ is an increasing function of the absolute value of the slope of lines through the point $(0,1)$. And since the absolute value of the slope of l_1 is less than or equal to the absolute value of the slope of l_2, we get that $\pi_1(\Psi_{\tau\tau'}(g_1,g_1')) \leq \pi_1(\Psi_{\tau\tau'}(g_2,g_2'))$. $\qquad\square$

Note. This last lemma is analogous to the fact in Section 4 that the map $\Phi_{\tau\tau'}$ is an increasing function and hence if $g_1 \leq g_2$, then $\Phi_{\tau\tau'}(g_1) \leq \Phi_{\tau\tau'}(g_2)$. This lemma about $\Psi_{\tau\tau'}$ preserving an order relation on its domain will be used in a way similar to how the fact that $\Phi_{\tau\tau'}$ is increasing was used in Section 4.

Lemma 6.10. *The map $h(g, g') = (g', g)$ is a conjugacy between $\Psi_{\tau\tau'}$ and $\Psi_{\tau'\tau}$.*

Proof. This fact is pretty clear if we go back to the geometric process that $\Psi_{\tau\tau'}$ was derived from. Applying the geometric process to the intervals $[0, 1 - g]$ with thickness τ and $[g', 1]$ with thickness τ' is equivalent to applying the geometric process to the intervals $[0, 1 - g']$ with thickness τ' and $[g, 1]$ with thickness τ.   □

7. The Dynamics of $\Psi_{\tau\tau'}$

In this section we will prove the following theorem about the dynamics of the two parameter family of piecewise differentiable maps $\Psi_{\tau\tau'}$ that was defined in the last section. This theorem about the dynamics of $\Psi_{\tau\tau'}$ is analogous to Proposition 4.3, which was about the dynamics of $\Phi_{\tau\tau'}$.

Proposition 7.1. *Assume that* $\tau\tau' > 1$.
  (1) *The point* $(0,0)$ *is a hyperbolic sink for* $\Psi_{\tau\tau'}$.
  (2) *If* τ *and* τ' *also satisfy the inequalities*

$$\tau \le \frac{\tau'^2 + 3\tau' + 1}{\tau'^2} \qquad \text{and} \qquad \tau' \le \frac{\tau^2 + 3\tau + 1}{\tau^2},$$

  then the point

$$\left(\frac{1}{1+\tau}, \frac{1}{1+\tau'} \right)$$

  is a fixed point of $\Psi_{\tau\tau'}$.
  (3) *If* τ *and* τ' *satisfy the inequalities*

$$\tau \le \frac{(1+2\tau')^2}{\tau'^3} \qquad \text{and} \qquad \tau' \le \frac{(1+2\tau)^2}{\tau^3},$$

  then the point

$$\left(\frac{\sqrt{\tau\tau'} - 1}{\sqrt{\tau\tau'} + \tau}, \frac{\sqrt{\tau\tau'} - 1}{\sqrt{\tau\tau'} + \tau'} \right)$$

  is a fixed point of $\Psi_{\tau\tau'}$.
  (4) *If* τ *and* τ' *do not satisfy the inequalities in the last two items (i.e., if* $(\tau, \tau') \in \Lambda_1$), *then* $(0,0)$ *is the only fixed point of* $\Psi_{\tau\tau'}$ *and the basin of attraction for* $(0,0)$ *is the whole domain of* $\Psi_{\tau\tau'}$.

Lemma 7.2. *If* $\tau\tau' > 1$, *then* $(0,0)$ *is a hyperbolic sink of* $\Psi_{\tau\tau'}$.

Proof. It is easy to see that $(0,0)$ is always a fixed point of $\Psi_{\tau\tau'}$. The derivative of $\Psi_{\tau\tau'}$ at $(0,0)$ is given by

$$d\Psi_{\tau\tau'}\Big|_{(0,0)} = \left(\begin{array}{cc} \dfrac{g'}{\tau\,(1-g-g')^2} & \dfrac{(1-g-g')+g'}{\tau\,(1-g-g')^2} \\[2ex] \dfrac{(1-g-g')+g}{\tau'\,(1-g-g')^2} & \dfrac{g}{\tau'\,(1-g-g')^2} \end{array} \right)\Bigg|_{(0,0)} = \left(\begin{array}{cc} 0 & \dfrac{1}{\tau} \\[2ex] \dfrac{1}{\tau'} & 0 \end{array} \right)$$

The eigenvalues of the derivative at $(0,0)$ are $\pm 1/\sqrt{\tau\tau'}$ which, by our hypothesis that $\tau\tau' > 1$, have absolute value strictly less than 1. $\qquad\square$

Next we will prove the second part of Proposition 1.

Lemma 7.3. *Assume that τ and τ' satisfy $\tau\tau' > 1$ and also the inequalities*

$$\tau \leq \frac{\tau'^2 + 3\tau' + 1}{\tau'^2} \qquad \text{and} \qquad \tau' \leq \frac{\tau^2 + 3\tau + 1}{\tau^2}.$$

Then the point

$$\left(\frac{1}{1+\tau}, \frac{1}{1+\tau'}\right)$$

is a fixed point of $\Psi_{\tau\tau'}$.

Proof. Notice that the point given in the statement of the lemma is the image of region (iv). Since $\Psi_{\tau\tau'}$ is a constant function in region (iv), if the image of region (iv) is contained in region (iv), then the image of region (iv) will be a fixed point (and this fixed point will be a hyperbolic sink if it is in the interior of region (iv)). So we need to solve the two inequalities

$$\frac{1}{1+\tau'} \geq \left(1 - \frac{1}{1+\tau}\right)\frac{\tau}{1+2\tau} \qquad \text{and} \qquad \frac{1}{1+\tau} \geq \left(1 - \frac{1}{1+\tau'}\right)\frac{\tau'}{1+2\tau'}.$$

The left hand inequality says that the image of region (iv) extends above the bottom edge of region (iv) (i.e., the boundary between regions (ii) and (iv)). The right hand inequality says that the image of region (iv) extends to the right of the left edge of region (iv) (i.e., the boundary between regions (iii) and (iv)). It is straightforward to show that the inequalities given above are equivalent to the ones in the statement of the lemma (the left hand one above is equivalent to the right hand one in the statement of the lemma). $\square$

In order to prove part (3) of Proposition 1, we need to analyze the definition of $\Psi_{\tau\tau'}$ in region (i). Let $\overline{\Psi}_{\tau\tau'}$ denote the function

$$\overline{\Psi}_{\tau\tau'}(g,g') = \left(\frac{g'}{\tau(1-g-g')}, \frac{g}{\tau'(1-g-g')}\right)$$

with domain $\{(g,g') \mid g, g' \geq 0,\ g + g' < 1\}$. The range of $\overline{\Psi}_{\tau\tau'}$ will be the whole positive quadrant $\{(g,g') \mid g, g' \geq 0\}$. Notice that what we have done here is to take the definition of $\Psi_{\tau\tau'}$ in region (i) and extend that definition to the whole domain of $\Psi_{\tau\tau'}$.

Now we want to look at the geometry of the map $\overline{\Psi}_{\tau\tau'}$.

Lemma 7.4.

(1) $\overline{\Psi}_{\tau\tau'}$ *maps a line through the origin with slope $m > 0$ to the line through the origin with slope $\tau/m\tau'$.*

(2) $\overline{\Psi}_{\tau\tau'}^2$ *leaves lines through the origin with positive slope invariant.*

(3) $\overline{\Psi}_{\tau\tau'}$ *leaves the line through the origin with slope $m = \sqrt{\tau/\tau'}$ invariant.*

Proof. Let $g' = mg$. Then

$$\overline{\Psi}_{\tau\tau'}(g, mg) = \left(\frac{mg}{\tau(1-g-mg)}, \frac{g}{\tau'(1-g-mg)}\right)$$

and it is easy to see that the second coordinate of $\overline{\Psi}_{\tau\tau'}(g, mg)$ is equal to $\tau/m\tau'$ times the first coordinate of $\overline{\Psi}_{\tau\tau'}(g, mg)$.

A line through the origin with slope $\tau/m\tau'$ will be mapped by $\overline{\Psi}_{\tau\tau'}$ to a line through the origin with slope

$$\frac{\tau}{(\tau/m\tau')\tau'} = m$$

so we see that $\overline{\Psi}_{\tau\tau'}^2$ leaves lines through the origin invariant.

To get a line through the origin with positive slope that is invariant under the map $\overline{\Psi}_{\tau\tau'}$, we need to solve the equation

$$\frac{\tau}{m\tau'} = m$$

which has as its positive solution $m = \sqrt{\tau/\tau'}$. Notice that the line through the origin with this slope is also the eigenspace for the eigenvalue $1/\sqrt{\tau\tau'}$ of the derivative of $\overline{\Psi}_{\tau\tau'}$ at the origin. $\qquad\square$

To find the fixed points of $\overline{\Psi}_{\tau\tau'}$, it helps to first compute an equation for $\overline{\Psi}_{\tau\tau'}^2$.

Lemma 7.5. *The map $\overline{\Psi}_{\tau\tau'}^2$ has the following equation*

$$\overline{\Psi}_{\tau\tau'}^2(g, g') = \frac{1}{\tau\tau'(1 - g - g') - g'\tau' - g\tau}(g, g').$$

The domain of $\overline{\Psi}_{\tau\tau'}^2$ is the subset of the domain of $\overline{\Psi}_{\tau\tau'}$ that satisfies the inequality

$$g' < \frac{\tau\tau'}{\tau\tau' + \tau'} - g\frac{\tau\tau' + \tau}{\tau\tau' + \tau'}.$$

Proof. The point $\overline{\Psi}_{\tau\tau'}^2(g, g')$ has the following coordinates

$$\left(\frac{\dfrac{g}{\tau\tau'(1 - g - g')}}{1 - \dfrac{g'}{\tau(1 - g - g')} - \dfrac{g}{\tau'(1 - g - g')}}, \frac{\dfrac{g'}{\tau\tau'(1 - g - g')}}{1 - \dfrac{g'}{\tau(1 - g - g')} - \dfrac{g}{\tau'(1 - g - g')}} \right).$$

Simplifying this expression will give the formula for $\overline{\Psi}_{\tau\tau'}^2$ given in the lemma.

To get the domain of $\overline{\Psi}_{\tau\tau'}^2$ we need to solve the inequality

$$\tau\tau'(1 - g - g') - g'\tau' - g\tau > 0$$

and a straightforward calculation gives the inequality stated in the lemma. $\qquad\square$

This equation for $\overline{\Psi}_{\tau\tau'}^2$ shows once again that $\overline{\Psi}_{\tau\tau'}^2$ leaves lines through the origin invariant.

Notice that if we let $g = 0$ then

$$\pi_2\left(\overline{\Psi}_{\tau\tau'}^2(0, g') \right) = \overline{\Phi}_{\tau\tau'}(g')$$

(where $\overline{\Phi}_{\tau\tau'}$ is defined in Section 4) which is pretty reasonable if you think about the geometric processes these maps were derived from. However, this equality does not extend to the maps $\Psi_{\tau\tau'}^2$ and $\Phi_{\tau\tau'}$ (probably because $g = 0$ does not have a very good geometric significance for the geometric process that $\Psi_{\tau\tau'}$ is derived from).

Lemma 7.6. *The line*

$$g' = \frac{\tau\tau' - 1}{\tau\tau' + \tau'} - g\frac{\tau\tau' + \tau}{\tau\tau' + \tau'}$$

is an invariant line for $\Psi_{\tau\tau'}$ and is made up entirely of period two points.

Proof. From the equation for $\overline{\Psi}_{\tau\tau'}^2$ derived in the last lemma we see that fixed points for $\overline{\Psi}_{\tau\tau'}^2$, or equivalently, period two points of $\overline{\Psi}_{\tau\tau'}$, can be found by solving the equation

$$1 = \tau\tau'(1 - g - g') - g'\tau' - g\tau$$

which is an equivalent expression for the line given in the lemma.

A period two orbit for $\overline{\Psi}_{\tau\tau'}$ will consist of the intersection of the line $g' = mg$ with this period two line and the intersection of the line $g' = (\tau/m\tau')g$ with the period two line. $\qquad\square$

Lemma 7.7. *The point*

$$\left(\frac{\sqrt{\tau\tau'} - 1}{\sqrt{\tau\tau'} + \tau}, \frac{\sqrt{\tau\tau'} - 1}{\sqrt{\tau\tau'} + \tau'}\right)$$

is a fixed point for $\overline{\Psi}_{\tau\tau'}$.

Proof. To get this fixed point we need to find the intersection of the period two line of $\overline{\Psi}_{\tau\tau'}$ with the invariant line through the origin which has slope $m = \sqrt{\tau/\tau'}$. Another straightforward calculation will show that the point given in the lemma is the intersection of these two lines. $\qquad\square$

The following lemma describes the "dynamics" of the map $\overline{\Psi}_{\tau\tau'}$. It is not really accurate to talk about the dynamics of $\overline{\Psi}_{\tau\tau'}$ since the image of the domain overflows the domain, so for some initial values of (g, g'), the iterates $\overline{\Psi}_{\tau\tau'}^n(g, g')$ are not defined for all values of n.

Lemma 7.8. *The subset of the domain of $\overline{\Psi}_{\tau\tau'}$ that is below the line of period two points is the basin of attraction for the hyperbolic sink at the origin. Points in this set are attracted to the origin by $\overline{\Psi}_{\tau\tau'}^2$ along rays through the origin.*

Points that are in the subset of the domain of $\overline{\Psi}_{\tau\tau'}$ that is above the line of period two points are repelled by $\overline{\Psi}_{\tau\tau'}^2$ away from the origin along rays through the origin until they leave the domain of $\overline{\Psi}_{\tau\tau'}$.

Proof. In the equation for $\overline{\Psi}_{\tau\tau'}^2$ derived above, the expression

$$\frac{1}{\tau\tau'(1 - g - g') - g'\tau' - g\tau}$$

has value strictly less than one exactly for those (g, g') that lie below the line of period two points and the value of this expression is strictly greater than one for those (g, g') that are above the line of period two points (see the proof of Lemma 6). $\qquad\square$

The fixed point of $\overline{\Psi}_{\tau\tau'}$ may or may not be a fixed point of $\Psi_{\tau\tau'}$. It depends on whether or not the fixed point of $\overline{\Psi}_{\tau\tau'}$ lies in region (i). The next lemma determines when the fixed point of $\overline{\Psi}_{\tau\tau'}$ lies in this domain of definition of $\Psi_{\tau\tau'}$.

Lemma 7.9. *If τ and τ' satisfy $\tau\tau' > 1$ and also the inequalities*

$$\tau \le \frac{(1+2\tau')^2}{\tau'^3} \qquad and \qquad \tau' \le \frac{(1+2\tau)^2}{\tau^3},$$

then the point

$$g = \frac{\sqrt{\tau\tau'}-1}{\sqrt{\tau\tau'}+\tau} \qquad and \qquad g' = \frac{\sqrt{\tau\tau'}-1}{\sqrt{\tau\tau'}+\tau'}$$

is a fixed point of $\Psi_{\tau\tau'}$.

Proof. We need to solve the two inequalities

$$\frac{\sqrt{\tau\tau'}-1}{\sqrt{\tau\tau'}+\tau'} \le \left(1 - \frac{\sqrt{\tau\tau'}-1}{\sqrt{\tau\tau'}+\tau}\right)\frac{\tau}{1+2\tau}$$

and

$$\frac{\sqrt{\tau\tau'}-1}{\sqrt{\tau\tau'}+\tau} \le \left(1 - \frac{\sqrt{\tau\tau'}-1}{\sqrt{\tau\tau'}+\tau'}\right)\frac{\tau'}{1+2\tau'}.$$

The first inequality says that the fixed point of $\overline{\Psi}_{\tau\tau'}$ is below the top edge of region (i). The second inequality says that the fixed point of $\overline{\Psi}_{\tau\tau'}$ is to the left of the right edge of region (i). When both inequalities are satisfied, the fixed point of $\overline{\Psi}_{\tau\tau'}$ is in region (i) and therefore it is also a fixed point of $\Psi_{\tau\tau'}$. The first inequality above is equivalent to the right hand inequality in the statement of the lemma and the second inequality above is equivalent to the left hand inequality. $\square$

So far we have found all of the fixed points of $\Psi_{\tau\tau'}$ that lie in regions (i) or (iv). The next lemma states that if for some value of τ and τ' there is a fixed point of $\Psi_{\tau\tau'}$ in regions (ii) or (iii), then there must also be a fixed point in either region (i) or region (iv). In other words, the only parameter values τ and τ' for which $\Psi_{\tau\tau'}$ has a fixed point other than the origin are those parameter values calculated above for which $\Psi_{\tau\tau'}$ has a fixed point in either region (i) or region (iv). The proof of this lemma is a long, tedious algebraic calculation and it has been put in an appendix.

Lemma 7.10. *If $\Psi_{\tau\tau'}$ has a fixed point other than the origin for some parameter values τ and τ', then these parameter values must lie in the set*

$$\left\{(\tau,\tau') \,\middle|\, \tau \le \frac{\tau'^2+3\tau'+1}{\tau'^2} \text{ and } \tau' \le \frac{\tau^2+3\tau+1}{\tau^2}\right\}$$

$$\bigcup\left\{(\tau,\tau') \,\middle|\, \tau \le \frac{(1+2\tau')^2}{\tau'^3} \text{ and } \tau' \le \frac{(1+2\tau)^2}{\tau^3}\right\}.$$

Now we shall begin working on part 4 (the last part) of Proposition 1. We want to show that when the origin is the only fixed point of $\Psi_{\tau\tau'}$, its basin of attraction is the whole domain of $\Psi_{\tau\tau'}$.

Let p denote the point where the lines

$$g' = (1-g)\frac{\tau}{1+2\tau} \quad \text{and} \quad g = (1-g')\frac{\tau'}{1+2\tau'}$$

intersect, i.e., the upper right hand corner of region (i). Then

$$\Psi_{\tau\tau'}(p) = \left(\frac{1}{1+\tau}, \frac{1}{1+\tau'}\right)$$

i.e., the image of p is the upper right hand corner of the rectangle that is the image of the map $\Psi_{\tau\tau'}$. The next lemma says that the orbit of p acts as a kind of upper bound on the orbit of any point in the domain of $\Psi_{\tau\tau'}$.

Lemma 7.11. *If for some parameter values τ and τ' we have $\Psi_{\tau\tau'}^n(p) \to (0,0)$ as $n \to \infty$, then $\Psi_{\tau\tau'}^n(g,g') \to (0,0)$ as $n \to \infty$ for all points (g,g') in the domain of $\Psi_{\tau\tau'}$.*

Proof. Suppose that $\Psi_{\tau\tau'}^n(p) \to (0,0)$ as $n \to \infty$. Let (g,g') be an arbitrary point in the domain of $\Psi_{\tau\tau'}$. Since $\Psi_{\tau\tau'}(p)$ is the upper right hand corner of the rectangle that is the image of $\Psi_{\tau\tau'}$ and since $\Psi_{\tau\tau'}(g,g')$ must be in this rectangle, we have $\Psi_{\tau\tau'}(g,g') \le \Psi_{\tau\tau'}(p)$ (where we are using the order relation $(x_1,y_1) \le (x_2,y_2)$ if and only if $x_1 \le x_2$ and $y_1 \le y_2$). Then by Lemma 6.9 and induction we get

$$\Psi_{\tau\tau'}^n(g,g') \le \Psi_{\tau\tau'}^n(p)$$

for $n \ge 1$. So then $\Psi_{\tau\tau'}^n(g,g') \to (0,0)$ as $n \to \infty$. $\square$

Because of this last lemma, to prove part 4 of Proposition 1 we only need to show that if the origin is the only fixed point of $\Psi_{\tau\tau'}$, then $\Psi_{\tau\tau'}^n(p) \to (0,0)$ as $n \to \infty$.

Lemma 7.12. *If τ and τ' are such that $\Psi_{\tau\tau'}$ has only the origin as a fixed point, then*

$$\Psi_{\tau\tau'}^n(p) \to (0,0) \text{ as } n \to \infty.$$

Proof. Notice that $\Psi_{\tau\tau'}(p)$ cannot be in region (iv) else $\Psi_{\tau\tau'}(p)$ would be a fixed point. Suppose that $\Psi_{\tau\tau'}(p)$ is in region (i). Then $\Psi_{\tau\tau'}^2(p) = \overline{\Psi}_{\tau\tau'}^2(p)$ and so $\Psi_{\tau\tau'}^2(p)$ is on the same line through the origin as p. But to stay in the image of $\Psi_{\tau\tau'}$, the point $\Psi_{\tau\tau'}^2(p)$ must be closer to the origin than p. This implies that the point p is below the line of period two points of $\overline{\Psi}_{\tau\tau'}$ and so p is in the basin of attraction of the origin for $\overline{\Psi}_{\tau\tau'}$. But since $\overline{\Psi}_{\tau\tau'}$ and $\Psi_{\tau\tau'}$ agree in region (i), the point p will also be in the basin of attraction of the origin for $\Psi_{\tau\tau'}$.

Now suppose that $\Psi_{\tau\tau'}(p)$ is in the interior of region (iii). Let ℓ_1 denote the line through the origin that is invariant under $\overline{\Psi}_{\tau\tau'}$ (i.e., the line with slope $\sqrt{\tau/\tau'}$). Let q denote the intersection of ℓ_1 with the line $g' = (1-g)\frac{\tau}{1+2\tau}$ (i.e., the line that includes the top edge of

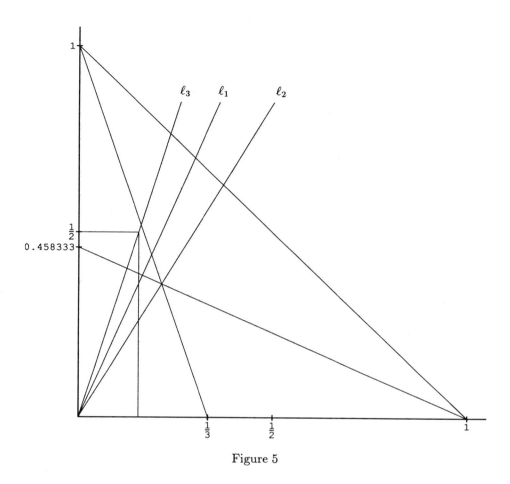

Figure 5

region (i)). Let ℓ_2 denote the line through the origin and the point p and let ℓ_3 denote the line through the origin and the point $\Psi_{\tau\tau'}(p)$. See Figure 5 (where $\tau = 5.5$ and $\tau' = 1$).

Claim 1. The slope of ℓ_1 is strictly greater than the slope of ℓ_2

If ℓ_2 and ℓ_3 were the same line, then $\Psi_{\tau\tau'}(p)$ would have to be in region (i), contradicting our assumption that $\Psi_{\tau\tau'}(p)$ is in the interior of region (iii). Since ℓ_2 and ℓ_3 are images of each other under $\overline{\Psi}_{\tau\tau'}$, they must lie on opposite sides of ℓ_1 (because of the geometry of the map $\overline{\Psi}_{\tau\tau'}$). Since $\Psi_{\tau\tau'}(p)$ is in region (iii), the line ℓ_3 must be above the line ℓ_2, and so the line ℓ_1 must also be above the line ℓ_2, i.e., the slope of ℓ_1 is strictly greater than the slope of ℓ_2.

Claim 2.

$$\frac{1}{1+\tau} < \pi_1(q) < \pi_1(p).$$

Since the slope of ℓ_1 is strictly greater than the slope of ℓ_2, we clearly must have $\pi_1(q) < \pi_1(p)$. Notice that $\Psi_{\tau\tau'}(q)$ must be the intersection of the line $g = 1/(1+\tau)$ with the invariant line because $\Psi_{\tau\tau'}(q)$ must stay on the invariant line and the line $g = 1/(1+\tau)$ is the image of the top edge of region (i) upon which lies q.

Suppose that $\pi_1(q) = 1/(1+\tau)$. But then $q = \Psi_{\tau\tau'}(q)$ is a fixed point, contradicting that the origin is the only fixed point. Suppose that $\pi_1(q) < 1/(1+\tau)$. But then $\Psi_{\tau\tau'}(q)$ is farther away from the origin than q which implies that the fixed point of $\overline{\Psi}_{\tau\tau'}$ lies closer to the origin on ℓ_1 than q. But then the fixed point of $\overline{\Psi}_{\tau\tau'}$ is also a fixed point of $\Psi_{\tau\tau'}$, again contradicting that the origin is the only fixed point of $\Psi_{\tau\tau'}$. So it must be that $\pi_1(q) > 1/(1+\tau)$. This finishes the proof of the last claim.

Claim 3. The segment of the vertical line $g = 1/(1+\tau)$ that is contained in region (i) is in the basin of attraction of the origin.

Claim 2 implies that the point q lies closer to the origin on the line ℓ_1 than the fixed point of $\overline{\Psi}_{\tau\tau'}$. So q lies below the line of period two points of $\overline{\Psi}_{\tau\tau'}$. The line of period two points must intersect the line $g' = (1-g)\frac{\tau}{1+2\tau}$ to the right of q (since the line of period two points intersects the g-axis to the left of 1) so the line of period two points intersects the vertical line $g = 1/(1+\tau)$ at a point in region (iii). So the segment of $g = 1/(1+\tau)$ that is in region (i) is below the line of period two points and hence in the basin of attraction of the origin for the map $\overline{\Psi}_{\tau\tau'}$. To show that this segment is also in the basin of attraction of the origin for the map $\Psi_{\tau\tau'}$ we need to show that the image of this segment is contained in region (i). Because of Lemma 6.9, it suffices to show that the top point of this segment has its image under $\Psi_{\tau\tau'}$ contained in region (i). This top point is on the line $g' = (1-g)\frac{\tau}{1+2\tau}$ to the left of q. So its image is below the image of q. And $\Psi_{\tau\tau'}(q)$ is contained in region (i), so the image of the top of the segment is also contained in region (i). So the segment of $g = 1/(1+\tau)$ contained in region (i) is in the basin of attraction of the origin for $\Psi_{\tau\tau'}$.

Claim 4. The iterates $\Psi_{\tau\tau'}^n(p)$ move strictly down the vertical line $g = 1/(1+\tau)$ until one lands on the segment of $g = 1/(1+\tau)$ that is in region (i).

Let r be a point in region (iii) that is on the line $g = 1/(1+\tau)$ and below the point $\Psi_{\tau\tau'}(p)$. Define the *shadow* of r to be the intersection of the line through $(0, 1)$ and r with the line $g' = (1-g)\frac{\tau}{1+2\tau}$. Notice that the shadow of r will be on the boundary between regions (i) and (iii). Also, the image of r and the image of its shadow under $\Psi_{\tau\tau'}$ are the same and this image point must again lie on the line $g = 1/(1+\tau)$ (see Corollary 6.4 and Lemma 6.8). Also, as the point r moves down the line $g = 1/(1+\tau)$, its shadow moves strictly to the left on the line $g' = (1-g)\frac{\tau}{1+2\tau}$ and so the image $\Psi_{\tau\tau'}(r)$ moves strictly down the line $g = 1/(1+\tau)$.

We know that the point $\Psi_{\tau\tau'}^2(p)$ is strictly lower on $g = 1/(1+\tau)$ than $\Psi_{\tau\tau'}(p)$. Suppose that $\Psi_{\tau\tau'}^{n+1}(p)$ is strictly lower on $g = 1/(1+\tau)$ than $\Psi_{\tau\tau'}^n(p)$ and still in region (iii). Then the shadow of $\Psi_{\tau\tau'}^{n+1}(p)$ is strictly to the left of the shadow of $\Psi_{\tau\tau'}^n(p)$. So $\Psi_{\tau\tau'}^{n+2}(p)$ is strictly lower on $g = 1/(1+\tau)$ than $\Psi_{\tau\tau'}^{n+1}(p)$. If $\Psi_{\tau\tau'}^{n+2}(p)$ is in region (i), then we have proved Claim 4. Otherwise, we repeat the induction to get a sequence of points $\{\Psi_{\tau\tau'}^n(p)\}$ moving strictly down the line $g = 1/(1+\tau)$. If this sequence is always in region (iii), it must have an accumulation point. But this accumulation point must be a fixed point of $\Psi_{\tau\tau'}$, which contradicts our assumption that the origin is the only fixed point of $\Psi_{\tau\tau'}$. So some point $\Psi_{\tau\tau'}^n(p)$ eventually lands on the segment of $g = 1/(1+\tau)$ that is in region (i). Which proves Claim 4.

Claims 3 and 4 together imply that p is in the basin of attraction of the origin. This completes the proof of the lemma. □

The last thing to do in this section is to draw graphs of the regions in parameter space that define the dynamics of $\Psi_{\tau\tau'}$. First we shall graph the equations

$$\tau = \frac{\tau'^2 + 3\tau' + 1}{\tau'^2} \qquad \text{and} \qquad \tau' = \frac{\tau^2 + 3\tau + 1}{\tau^2}.$$

The first equation has $\tau = 1$ as a vertical asymptote and $\tau' = 0$ as a horizontal asymptote. The second equation has $\tau' = 1$ as a horizontal asymptote and $\tau = 0$ as a vertical asymptote. Their graphs are symmetric with respect to the line $\tau = \tau'$. Let's find where they cross this line. Setting $\tau = \tau'$ in either of the above two equations gives an equation equivalent to $0 = (\tau + 1)(\tau^2 - 2\tau - 1)$. So we see that these two equations intersect the main diagonal at the point $(\tau, \tau') = (1 + \sqrt{2}, 1 + \sqrt{2})$. The part of parameter space where $\Psi_{\tau\tau'}$ has a fixed point in region (iv) is given in Figure 6.

Now we shall graph the equations

$$\tau = \frac{(1 + 2\tau')^2}{\tau'^3} \qquad \text{and} \qquad \tau' = \frac{(1 + 2\tau)^2}{\tau^3}.$$

Both equations have $\tau' = 0$ as a horizontal asymptote and $\tau = 0$ as a vertical asymptote. These equations are also symmetric with respect to the line $\tau = \tau'$. To find where they cross this line, let $\tau = \tau'$ in either of the two equations. This gives an equation equivalent to $0 = (\tau + 1)^2(\tau^2 - 2\tau - 1)^2$. So we see that these two graphs also intersect the main diagonal at the point $(\tau, \tau') = (1 + \sqrt{2}, 1 + \sqrt{2})$. Before we can draw the graphs of these two equations, we need to determine which equation is above the other for $\tau > 1 + \sqrt{2}$ and for $\tau < 1 + \sqrt{2}$. The easiest way to do this is to just plot a couple of points, since we know that the equations only cross at one point. The part of parameter space where $\Psi_{\tau\tau'}$ has a fixed point in region (i) is shown if Figure 7.

Now we want to combine the last two graphs to get a picture of the set Λ_2. Since Λ_2 is symmetric with respect to the line $\tau = \tau'$, we will at first just look at the part of Λ_2 contained in the region where $\tau \geq \tau'$, i.e., the part of Λ_2 under the main diagonal. We need to determine the relationship between the graphs of

$$\tau = \frac{\tau'^2 + 3\tau' + 1}{\tau'^2} \qquad \text{and} \qquad \tau' = \frac{(1 + 2\tau)^2}{\tau^3}.$$

We know that they intersect on the diagonal $\tau = \tau'$ at the point $(1 + \sqrt{2}, 1 + \sqrt{2})$. Now we shall determine if they intersect anywhere else. First let's invert the first function. Convert the first function to the equation

$$0 = (1 - \tau)\tau'^2 + 3\tau' + 1$$

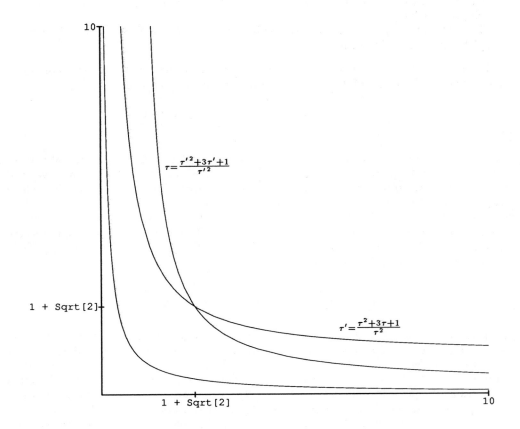

$$\tau = \frac{\tau'^2 + 3\tau' + 1}{\tau'^2}$$

$$\tau' = \frac{\tau^2 + 3\tau + 1}{\tau^2}$$

Figure 6

and then use the quadratic formula to solve for τ' as a function of τ to get

$$\tau' = \frac{-3 - \sqrt{9 - 4(1 - \tau)}}{2(1 - \tau)}$$
$$= \frac{3 + \sqrt{5 + 4\tau}}{2(\tau - 1)}.$$

Now we need to solve

$$\frac{(1 + 2\tau)^2}{\tau^3} = \frac{3 + \sqrt{5 + 4\tau}}{2(\tau - 1)}.$$

After a lot of multiplying out and squaring and all that, we get the polynomial equation

$$0 = 1 + 6\tau + 9\tau^2 - 5\tau^3 - 15\tau^4 + 5\tau^6 - \tau^7$$

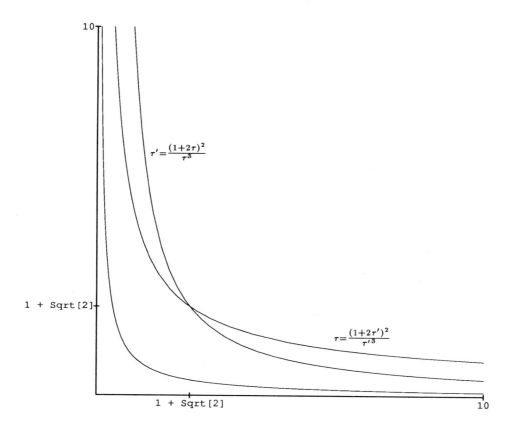

Figure 7

which factors into

$$0 = (1 - \tau)(1 + \tau)(\tau^2 - 2\tau - 1)(\tau^3 - 3\tau^2 - 4\tau - 1).$$

So other possible intersections of these two equations are determined by the roots of the polynomial

$$p(\tau) = \tau^3 - 3\tau^2 - 4\tau - 1.$$

It is not hard to show that this polynomial has only one positive real root. Denote this root by τ_0. The value of this root is approximately $\tau_0 = 4.04892$. For $\tau \in (1 + \sqrt{2}, \tau_0)$, the graph of the equation

$$\tau' = \frac{(1 + 2\tau)^2}{\tau^3}$$

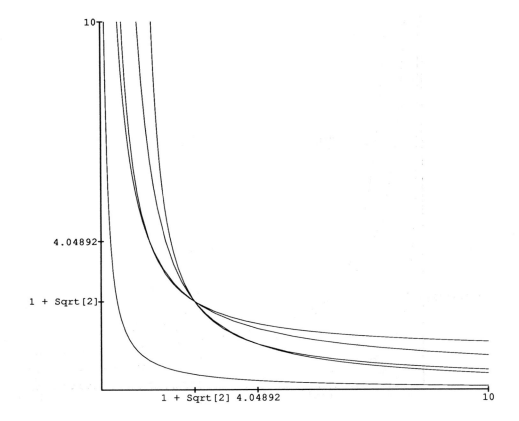

Figure 8

is above the graph of

$$\tau = \frac{\tau'^2 + 3\tau' + 1}{\tau'^2}$$

and the opposite is true for $\tau > \tau_0$. The combined graph of all four equations used in the definitions of Λ_1 and Λ_2 is shown in Figure 8 (in other words, the combination of the last two graphs). The graph of the sets Λ_1 and Λ_2 in parameter space is shown in Figure 1.

8. Results About The Geometric Process

In this section we will use the map $\Psi_{\tau\tau'}$ and what we know about its dynamics to determine for which thicknesses the geometric process defined in Section 6 succeeds in constructing Cantor sets and for which thicknesses it fails. When the geometric process does succeed, we will say some things about the structure of the Cantor sets it defines.

Theorem 8.1. *If for some parameter values τ and τ' the map $\Psi_{\tau\tau'}$ has $(g, g') \neq (0,0)$ as a fixed point, then the geometric process with thicknesses τ and τ' and fundamental intervals $[0, 1-g]$ and $[g', 1]$ will succeed in defining two Cantor sets.*

If for some parameter values τ and τ' the map $\Psi_{\tau\tau'}$ has only the origin as a fixed point, then the geometric process with thicknesses τ and τ' will fail for any (g, g').

The main tool for proving this theorem is the following lemma, which shows how the map $\Psi_{\tau\tau'}$ is used to get information about the geometric process.

Lemma 8.2. *Choose parameter values τ and τ' with $\tau\tau' > 1$ and a point (g, g') in the domain of $\Psi_{\tau\tau'}$. Let $(g_n, g_n') = \Psi_{\tau\tau'}^n(g, g')$. Then for the geometric process with thicknesses τ and τ' and fundamental intervals $[0, 1-g]$ and $[g', 1]$, we have the following formula for the length of the interval $B_n \cup B_n'$ (recall that B_n and B_n' are the overlapping bridges of the gaps U_n and U_n').*

$$|B_n \cup B_n'| = \prod_{k=0}^{n-1} [1 - (g_k + g_k')]$$

Proof. The proof is by induction. For $n = 1$, we clearly have $|B_1 \cup B_1'| = 1 - g - g' = 1 - g_0 - g_0'$ (see the following picture).

Now assume that

$$|B_{n-1} \cup B_{n-1}'| = \prod_{k=0}^{n-2} [1 - (g_k + g_k')].$$

Then if we take the picture

and divide all the lengths by $\prod_{k=0}^{n-2} [1 - (g_k + g'_k)]$, i.e., by the length of $B_{n-1} \cup B'_{n-1}$, then we get the following picture.

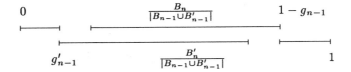

So then we see that

$$1 - g_{n-1} - g'_{n-1} = \frac{|B_n \cup B'_n|}{\prod_{k=0}^{n-2} [1 - (g_k + g'_k)]}. \qquad \square$$

Lemma 8.3. *If τ and τ' are such that $\Psi_{\tau\tau'}$ has only the origin as a fixed point, then the geometric process fails to define Cantor sets.*

Proof. The hypothesis implies that the basin of attraction for the hyperbolic sink at the origin is the whole domain of $\Psi_{\tau\tau'}$. So for any choice of (g, g'), the infinite series $\sum (g_k + g'_k)$ will converge. Therefore the infinite product $\prod [1 - (g_k + g'_k)]$ will converge to a nonzero positive number. But this implies that the lengths of the intervals B_n and B'_n do not go to zero as $n \to \infty$. So the geometric process fails to define two Cantor sets. $\qquad \square$

Lemma 8.4. *If τ and τ' are such that $\Psi_{\tau\tau'}$ has a fixed point other than the origin, and if $(g, g') > 0$ is chosen outside of the basin of attraction of the origin, then the geometric process with thicknesses τ and τ' and fundamental intervals $[0, 1-g]$ and $[g', 1]$ will succeed in defining two Cantor sets.*

Proof. If (g, g') is not in the basin of attraction of the origin, then the infinite series $\sum g_k$ and $\sum g'_k$ both diverge (it is not possible for (g_n, g'_n) to tend towards one of the axes other than at the origin). So then the infinite product $\prod [1 - (g_k + g'_k)]$ will have 0 as its limit. So the lengths of the intervals B_n and B'_n go to zero and the geometric process succeeds in defining two Cantor sets. $\qquad \square$

Note. If (g, g') is chosen to be a fixed point of $\Psi_{\tau\tau'}$, then the geometric process defines a self-similar construction. If (g, g') is chosen to be a period two point of $\Psi_{\tau\tau'}$, then the geometric process is self-similar at every other step. If (g, g') is chosen to be in the basin of attraction of a fixed point, then the geometric process will tend towards self-similarity, or in other words, it will be eventually self-similar.

Now we shall give a formula for the value of the unique overlapped point in $\Gamma \cap \Gamma'$ in the case where (g, g') is a fixed point of $\Psi_{\tau\tau'}$.

Lemma 8.5. *If the point $(g, g') \neq (0, 0)$ is a fixed point of the map $\Psi_{\tau\tau'}$, then the unique overlapped point in $\Gamma \cap \Gamma'$ has the value*

$$\frac{1}{1 + \left(\dfrac{1 - g'}{1 - g}\right)}.$$

So in particular if $\tau\tau' > 1$ and also

$$\tau \le \frac{\tau'^2 + 3\tau' + 1}{\tau'^2} \qquad \text{and} \qquad \tau' \le \frac{\tau^2 + 3\tau + 1}{\tau^2}$$

and if we let

$$(g, g') = \left(\frac{1}{1+\tau}, \frac{1}{1+\tau'}\right)$$

then the unique overlapped point in $\Gamma \cap \Gamma'$ has the value

$$\frac{1}{1 + \dfrac{\tau'}{\tau}\dfrac{1+\tau}{1+\tau'}}.$$

If

$$\tau \le \frac{(1+2\tau')^2}{\tau'^3} \qquad \text{and} \qquad \tau' \le \frac{(1+2\tau)^2}{\tau^3}$$

and if we let

$$(g, g') = \left(\frac{\sqrt{\tau\tau'}-1}{\sqrt{\tau\tau'}+\tau}, \frac{\sqrt{\tau\tau'}-1}{\sqrt{\tau\tau'}+\tau'}\right)$$

then the unique overlapped point in $\Gamma \cap \Gamma'$ has the value

$$\frac{1}{1 + \dfrac{\sqrt{\tau}}{\sqrt{\tau'}}\dfrac{\tau'+1}{\tau+1}}.$$

Proof. When (g, g') is a fixed point of $\Psi_{\tau\tau'}$, the geometric process defines a self-similar construction. The linear map T that maps the point 0 to $1 - g$ and the point 1 to g' is a scaling transformation for the pair of Cantor sets Γ and Γ'. In other words, T will map the interval B_n to the interval B_{n+1} and it will simultaneously map B'_n to B'_{n+1}. The unique overlapped point in $\Gamma \cap \Gamma'$ is defined by $\bigcap_{n=1}^{\infty} B_n$, i.e., the point must be in all the intervals B_n. The only point that can be in every B_n is a fixed point of the scaling map T. So we need to compute the fixed point of T.

The map T has the formula $T(x) = (g + g' - 1)x + (1 - g)$. To compute its fixed point we need to solve $x = (g + g' - 1)x + (1 - g)$ which has the solution

$$x = \frac{1}{1 + \left(\dfrac{1-g'}{1-g}\right)}.$$

The particular cases given in the statement of the lemma come from plugging into this formula. $\square$

Notice that when $g = g'$, the overlapped point is $x = 1/2$. So in particular, when $\tau = \tau'$ the overlapped point is $1/2$ for either choice of the fixed point of $\Psi_{\tau\tau'}$. When $\tau = \tau'$ it is tempting to conclude that everything is symmetric and that the Cantor sets Γ and Γ' defined by the geometric process are middle-α Cantor sets. But this need not be true as we will now see.

If we look back at the definition of $\Psi_{\tau\tau'}$ in Section 6, the line $g' = (1-g)\tau/(1+2\tau)$ defines when the gap U_1 is centered in the middle of the interval $[0, 1-g]$ and the line $g = (1-g')\tau'/(1+2\tau')$ defines when the gap U_1' is centered in the middle of the interval $[g', 1]$. If (g, g') is a fixed point of $\Psi_{\tau\tau'}$, then Γ is a middle-α Cantor set if (g, g') is on $g' = (1-g)\tau/(1+2\tau)$ and Γ' is a middle-α Cantor set if (g, g') is on $g = (1-g')\tau'/(1+2\tau')$. So the only way that the geometric process can define a middle-α Cantor set is when a fixed point of $\Psi_{\tau\tau'}$ is on the boundary of regions (i) or (iv). The conditions on τ and τ' so that a fixed point (g, g') of $\Psi_{\tau\tau'}$ is on the boundary of (i) or (iv) are the equations that define the boundaries of the sets Λ_1 and Λ_2. In particular, the only way that Γ and Γ' can both be middle-α Cantor sets is if $(\tau, \tau') = (1 + \sqrt{2}, 1 + \sqrt{2})$.

Near the end of Section 2 we looked at the idea of removing a gap U from a closed interval $[a, b]$ such that the gap determines a given thickness τ and so that the center of the gap is at the point $(1-\theta)a + \theta b$ for some $\theta \in (0, 1)$. A middle-α Cantor set is one in which every gap has $\theta = 1/2$ (when the gaps are removed in order of decreasing size). If (g, g') is a fixed point of $\Psi_{\tau\tau'}$, then the self-similarity of Γ and Γ' can be expressed by saying that there is a constant value of θ for every gap U_n and another constant θ' for every gap U_n'. (Actually, the gaps U_n are alternately centered by θ and $1 - \theta$. Similarly for the U_n'.) In the next lemma we will compute these values of θ and θ' for fixed points of $\Psi_{\tau\tau'}$.

Lemma 8.6. *If (g, g') is a fixed point of $\Psi_{\tau\tau'}$, then*

$$\theta = \begin{cases} \dfrac{g'}{1-g} \dfrac{1+2\tau}{2\tau} & \text{if } g' \leq (1-g)\dfrac{\tau}{1+2\tau} \\[3mm] 1 - \dfrac{1-g-g'}{1-g} \dfrac{1+2\tau}{2(1+\tau)} & \text{if } g' \geq (1-g)\dfrac{\tau}{1+2\tau} \end{cases}$$

and

$$\theta' = \begin{cases} 1 - \dfrac{g}{1-g'} \dfrac{1+2\tau'}{2\tau'} & \text{if } g \leq (1-g')\dfrac{\tau'}{1+2\tau'} \\[3mm] \dfrac{1-g-g'}{1-g'} \dfrac{1+2\tau'}{2(1+\tau')} & \text{if } g \geq (1-g')\dfrac{\tau'}{1+2\tau'} \end{cases}$$

Proof. Formula (1) from Section 2 (which is between Lemmas 2.8 and 2.9) states that when $\theta \leq 1/2$,

$$|U| = \frac{2\theta}{1+2\tau}(b - a).$$

So for any $\theta \in (0, 1)$ we have

$$|U| = \begin{cases} \dfrac{2\theta}{1+2\tau}(b-a) & \text{if } \theta \leq 1/2 \\[3mm] \dfrac{2(1-\theta)}{1+2\tau}(b-a) & \text{if } \theta \geq 1/2 \end{cases}$$

and so, by solving for θ, we have

$$\theta = \begin{cases} |U|\dfrac{1+2\tau}{2(b-a)} & \text{if the left hand bridge of } U \text{ is shorter} \\[3mm] 1 - |U|\dfrac{1+2\tau}{2(b-a)} & \text{if the right hand bridge of } U \text{ is shorter.} \end{cases}$$

In the case of the gap U_1 determined by the fixed point we get

$$\theta = \begin{cases} |U_1|\dfrac{1+2\tau}{2(1-g)} & \text{if } g' \leq (1-g)\dfrac{\tau}{1+2\tau} \\[3mm] 1 - |U_1|\dfrac{1+2\tau}{2(1-g)} & \text{if } g' \geq (1-g)\dfrac{\tau}{1+2\tau} \end{cases}$$

which, when we substitute in the expressions for $|U_1|$ from Section 6, becomes

$$\theta = \begin{cases} \dfrac{g'}{\tau}\dfrac{1+2\tau}{2(1-g)} & \text{if } g' \leq (1-g)\dfrac{\tau}{1+2\tau} \\[3mm] 1 - \dfrac{1-g-g'}{1+\tau}\dfrac{1+2\tau}{2(1-g)} & \text{if } g' \geq (1-g)\dfrac{\tau}{1+2\tau} \end{cases}$$

which is the first formula given in the lemma. The other formula given in the lemma is derived in the same way, but keeping in mind that the right hand bridge of U_1' is the shorter bridge when g is small. $\qquad\square$

So in particular, for the fixed point

$$\left(\frac{1}{1+\tau}, \frac{1}{1+\tau'} \right)$$

the values for θ and θ' are

$$\theta = 1 - \frac{\tau\tau'-1}{(1+\tau)(1+\tau')}\frac{1+2\tau}{2\tau} \qquad \text{and} \qquad \theta' = \frac{\tau\tau'-1}{(1+\tau)(1+\tau')}\frac{1+2\tau'}{2\tau'}.$$

For the fixed point

$$\left(\frac{\sqrt{\tau\tau'}-1}{\sqrt{\tau\tau'}+\tau}, \frac{\sqrt{\tau\tau'}-1}{\sqrt{\tau\tau'}+\tau'} \right)$$

the values of θ and θ' are

$$\theta = \frac{\sqrt{\tau}}{\sqrt{\tau'}}\frac{\sqrt{\tau\tau'}-1}{\tau+1}\left(\frac{1+2\tau}{2\tau}\right) \qquad \text{and} \qquad \theta' = 1 - \frac{\sqrt{\tau'}}{\sqrt{\tau}}\frac{\sqrt{\tau\tau'}-1}{\tau'+1}\left(\frac{1+2\tau'}{2\tau'}\right).$$

Lemma 8.7. *If (g, g') is a fixed point of $\Psi_{\tau\tau'}$, then the geometric process can be redefined so that it constructs affine Cantor sets.*

Proof. The definition of an affine Cantor set is given at the end of Section 2.

Remove gaps U_1, U_1', U_2 and U_2' from the intervals $[0, 1-g]$ and $[g', 1]$ as in the geometric process but do not do anything yet in the nonintersecting bridges. Now define an affine Cantor set in the interval $[0, 1-g]$ using the three bridges L_1, L_2 and R_2 as the template. Define an affine Cantor set in $[g', 1]$ using the three bridges R_1', R_2', and L_2' as the template. The affine Cantor sets will have the same gaps U_n and U_n' as the geometric process. The only difference will be in the definition of the Cantor subsets in the nonintersecting bridges. $\square$

9. Proof Of Theorem 6.1(1)

In this section we will prove part (1) of Theorem 6.1.

Theorem 9.1. *Suppose that Γ and Γ' are Cantor sets with fundamental intervals $[0, 1-g]$ and $[g', 1]$ with $g + g' < 1$ and that Γ and Γ' have thicknesses τ and τ'. Also suppose that there is an $x \in \Gamma \cap \Gamma'$ where x is an overlapped point of the first kind. Let $\mathcal{U}$ and $\mathcal{U}'$ be any orderings of the gaps of Γ and Γ' by decreasing size. If $(\tau, \tau') \in \Lambda_1$, then for some n there will be two pairs of interleaved bridges in $C_n(\mathcal{U})$ and $C_n(\mathcal{U}')$.*

We will prove this theorem by proving its contrapositive, i.e., we will assume that for all n, $C_n(\mathcal{U})$ and $C_n(\mathcal{U}')$ have only one pair of interleaved bridges, the pair of bridges that always contain x in their interiors, and then we will show that $(\tau, \tau') \notin \Lambda_1$. The proof will be a long sequence of lemmas. Until further notice, we will make the following standing assumptions.

Assumptions. Let Γ and Γ' be Cantor sets that satisfy the following conditions.
   (1) The fundamental intervals of Γ and Γ' are $[0, 1-g]$ and $[g', 1]$ with $g + g' < 1$.
   (2) The thicknesses of Γ and Γ' are τ and τ' with $\tau\tau' > 1$.
   (3) There is a point x in $\Gamma \cap \Gamma'$ that is an overlapped point of the first kind.
   (4) For all n, $C_n(\mathcal{U})$ and $C_n(\mathcal{U}')$ have only one pair of interleaved bridges, the pair that contains x.

We want to see what conclusions we can draw about Γ and Γ' from these conditions. Notice that condition (4) implies that x is the only overlapped point in $\Gamma \cap \Gamma'$ even though there may be many other (nonoverlapped) points in the intersection.

Consider the point g'. Clearly it is in Γ'. It may or may not be in Γ. Suppose that g' is not in Γ. Then there is a gap of Γ, which we will call V_1, that contains g'. So we have the following picture.

On the other hand, if g' is in Γ, and hence in $\Gamma \cap \Gamma'$ also, then it cannot be an overlapped point and so it must be the left endpoint of a gap of Γ that we will again call V_1. So in this case we get the following picture.

In a similar way, we get a gap V_1' of Γ' that either contains the point $1 - g$ or has $1 - g$ as its right hand endpoint. Let A_1 and A_1' denote the overlapping bridges of V_1 and V_1'.

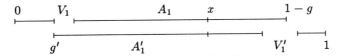

If we apply the argument of the last paragraph to the right endpoint of A_1' and to the left endpoint of A_1, then we get two more gaps V_2 and V_2'. Let A_2 and A_2' denote the overlapping pair of bridges for V_2 and V_2'.

In a manner analogous to the geometric process from Section 6, we can use the argument in the last couple of paragraphs to define two subsequences of gaps $\{V_n\}_{n=1}^\infty$ and $\{V_n'\}_{n=1}^\infty$ from Γ and Γ'. Let A_n and A_n' denote the overlapping pair of bridges that remains after V_n and V_n' have been removed. Notice that for all n, $x \in A_n \cap A_n'$.

The gaps $\{V_n\}_{n=1}^\infty$ and $\{V_n'\}_{n=1}^\infty$ are analogous to the gaps $\{U_n\}_{n=1}^\infty$ and $\{U_n'\}_{n=1}^\infty$ constructed by the geometric process with thicknesses τ and τ'. The bridges $\{A_n\}_{n=1}^\infty$ and $\{A_n'\}_{n=1}^\infty$ are analogous to the bridges $\{B_n\}_{n=1}^\infty$ and $\{B_n'\}_{n=1}^\infty$ from the geometric process. Now we will define two sequences of numbers $\{h_n\}_{n=0}^\infty$ and $\{h_n'\}_{n=0}^\infty$ that are analogous to the sequences $\{g_n\}_{n=0}^\infty$ and $\{g_n'\}_{n=0}^\infty$ which were defined by $(g_n, g_n') = \Psi_{\tau\tau'}^n(g, g')$.

Let $h_0 = g$ and $h_0' = g'$. Consider the following picture.

We will define α_n to be the distance between the endpoint of A_n' contained in V_n and the endpoint of V_n that is contained in A_n'. Similarly, we will define α_n' to be the distance between the endpoint A_n that is contained in V_n' and the endpoint of V_n' that is contained in A_n. Then for $n \geq 1$ we will define

$$h_n = \frac{\alpha_n}{|A_n \cup A_n'|} \quad \text{and} \quad h_n' = \frac{\alpha_n'}{|A_n \cup A_n'|}.$$

We now have the following formula, which is a generalization of Lemma 8.2, relating the numbers h_n and h_n' to the lengths of the intervals $A_n \cup A_n'$.

Lemma 9.2. *If the Cantor sets Γ and Γ' satisfy conditions (1)–(4), then for $n \geq 1$*

$$|A_n \cup A_n'| = \prod_{k=0}^{n-1} [1 - (h_k + h_k')].$$

Proof. The proof is by induction. Clearly we have

$$|A_1 \cup A_1'| = 1 - g - g' = 1 - h_0 - h_0'.$$

Now suppose that

$$|A_{n-1} \cup A_{n-1}'| = \prod_{k=0}^{n-2} [1 - (h_k + h_k')].$$

Now consider the following picture.

From this picture we see that

$$
\begin{aligned}
|A_n \cup A_n'| &= |A_{n-1} \cup A_{n-1}'| - \alpha_{n-1} - \alpha_{n-1}' \\
&= \left(1 - \frac{\alpha_{n-1}}{|A_{n-1} \cup A_{n-1}'|} - \frac{\alpha_{n-1}'}{|A_{n-1} \cup A_{n-1}'|}\right) |A_{n-1} \cup A_{n-1}'| \\
&= (1 - h_{n-1} - h_{n-1}') |A_{n-1} \cup A_{n-1}'| \\
&= [1 - (h_{n-1} + h_{n-1}')] \prod_{k=0}^{n-2} [1 - (h_k + h_k')] \\
&= \prod_{k=0}^{n-1} [1 - (h_k + h_k')].
\end{aligned}
$$

$\square$

It would be nice if it was true that the sequences of gaps $\{V_n\}_{n=1}^{\infty}$ and $\{V_n'\}_{n=1}^{\infty}$ are ordered by decreasing size, but it is not hard to find examples of Cantor sets Γ and Γ' that satisfy conditions (1)–(4) for which these sequences are not ordered this way (for example, the geometric process when g is much larger than g' and τ is smaller than τ'). However, the next three lemmas will prove a partial result about the orderings of these gaps.

Lemma 9.3. *The longest gap of Γ that intersects the interval $[g', 1]$ is either V_1 or V_2.*
The longest gap of Γ' that intersects the interval $[0, 1 - g]$ is either V_1' or V_2'.

Proof. We will prove the first sentence of the lemma. Let V be the longest gap of Γ that intersects $[g', 1]$. If V is not unique, then choose any one of the longest gaps of Γ that intersect $[g', 1]$. We will show that $V = V_1$ or $V = V_2$.

The gap V must lie either between the point 0 and the point x, or V must lie between the point x and the point $1 - g$. Suppose that V is between 0 and x. Either V contains the point g' or it does not. If V contains the point g', then by the definition of V_1, $V = V_1$. Suppose that V does not contain the point g'. Then V must lie between the gap V_1 and the point x. Remove from the interval $[0, 1 - g]$ all the gaps of Γ that have length greater than or equal to the length of V, including V, but do not remove V_1 (in the case where

$|V| = |V_1|$). Then there will remain a bridge of Γ that contains the point g' but does not contain the point x, and this bridge will be interleaved with the interval $[g', 1]$. By Lemma 2.2 and Newhouse's lemma, we can conclude that there is another overlapped point in $\Gamma \cap \Gamma'$ besides x. But this contradicts assumption (4). So we cannot have V between V_1 and x. So if V is between 0 and x, then $V = V_1$.

Now suppose that V is between x and $1 - g$. Remove from the interval $[0, 1 - g]$ all the gaps of Γ that have length strictly greater than the length of V and then remove the gap V itself. Then we will have a bridge from the right hand endpoint of V to the point $1 - g$. See the following picture (all the gaps of Γ that are strictly larger than V must lie to the left of g').

In order that the bridge to the right of V not be interleaved with $[g', 1]$, it must lie in a gap of Γ'. Let V' denote this gap of Γ' that contains V. Then we have the following picture.

It is clear that the gap V' is actually V_1' and that the gap V is actually V_2. So if V is between x and $1 - g$, then $V = V_2$.   □

Lemma 9.4. *We have either $|V_1| > |V_2|$ or $|V_1'| > |V_2'|$.*

Proof. Suppose that $|V_1| \leq |V_2|$ and $|V_1'| \leq |V_2'|$. Let R denote the right hand bridge of V_2 and let L denote the left hand bridge of V_2'. See the following picture.

Then we have

$$\tau\tau' \leq \frac{|R|}{|V_2|}\frac{|L|}{|V_2'|} \leq \frac{|V_1'|}{|V_2|}\frac{|V_1|}{|V_2'|} = \frac{|V_1|}{|V_2|}\frac{|V_1'|}{|V_2'|} \leq 1.$$

But this contradicts assumption (2).   □

Lemma 9.5. *Each of the sequences of gaps $\{V_{2n-1}\}_{n=1}^{\infty}$, $\{V_{2n-1}'\}_{n=1}^{\infty}$, $\{V_{2n}\}_{n=1}^{\infty}$ and $\{V_{2n}'\}_{n=1}^{\infty}$ is strictly ordered by decreasing size.*

Proof. By the last two lemmas, when we remove from $[0, 1 - g]$ the largest gap of Γ that intersects $[g', 1]$ and then we remove from $[g', 1]$ the largest gap of Γ' that intersects $[0, 1-g]$, we get one of the following three pairs of gaps, $\{V_1, V_1'\}$ or $\{V_1, V_2'\}$ or $\{V_2, V_1'\}$. So the largest such gaps have to come from the beginning of the sequences of gaps given in the statement of the lemma.

When these two largest gaps are removed, their overlapping bridges will still satisfy assumptions (2)–(4), which were the assumptions needed to prove the last two lemmas. So when the next pair of largest gaps is removed from these overlapping bridges, this pair too must come from the beginning of (what remains of) the four sequences of gaps defined above.

By iterating the argument in the last paragraph over and over again, we get that the even and odd subsequences of $\{V_n\}_{n=1}^{\infty}$ and $\{V_n'\}_{n=1}^{\infty}$ are strictly ordered by decreasing size. □

At this point it is worth noting something about the sequences of gaps $\{V_n\}_{n=1}^{\infty}$ and $\{V_n'\}_{n=1}^{\infty}$. In the ordering in which we have defined them, they are removed in a way analogous to the geometric process from Section 6, i.e., we remove them in pairs from opposite ends of the fundamental intervals. But the last three lemmas show that when these sequences are reordered by decreasing size, they may be removed in other ways. The two largest gaps can both come off of the same end of the fundamental intervals, either the right end or the left. When a pair of largest gaps are removed in one of these two other ways, this mimics the geometric process of Section 4. As we remove the gaps from the sequences $\{V_n\}_{n=1}^{\infty}$ and $\{V_n'\}_{n=1}^{\infty}$ (but in the order of decreasing size) we can end up bouncing in an arbitrary way between pairs of largest gaps that mimic the geometric process from Section 6 and pairs of largest gaps that mimic the geometric process of Section 4 (on different sides of the point x).

If the gaps $\{V_n\}_{n=1}^{\infty}$ and $\{V_n'\}_{n=1}^{\infty}$ were ordered by decreasing size, then Lemma 2.5 would tell us that in these orderings, each gap V_n would determine a thickness greater than or equal to τ and each gap V_n' would determine a thickness greater than or equal to τ'. The following lemma gives a useful partial result along these lines.

Lemma 9.6. *When the gaps $\{V_n\}_{n=1}^{\infty}$ are removed in that order, the ratio of the length of each V_n with the length of its nonoverlapping bridge is greater than or equal to τ.*
Similarly for the gaps $\{V_n'\}_{n=1}^{\infty}$ and the number τ'.

Note. For example, in the picture

$$\begin{array}{c} \quad\;\; L_n \qquad\qquad\quad V_n \;\; R_n \\ \vdash\!\!\!-\!\!\!-\!\!\!-\!\!\!-\!\!\!-\!\!\!-\!\!\!-\!\!\!\dashv \quad \vdash\!\!\!-\!\!\!\dashv \\ \\ \vdash\!\!\!-\!\!\!\dashv \;\; \vdash\!\!\!-\!\!\!-\!\!\!-\!\!\!-\!\!\!-\!\!\!-\!\!\!-\!\!\!-\!\!\!\dashv \\ L_n' \quad V_n' \qquad R_n' \end{array}$$

we have $|R_n|/|V_n| \geq \tau$ (and also $|L_n'|/|V_n'| \geq \tau'$). However, this ratio need not be the thickness determined by V_n in the ordering $\{V_n\}_{n=1}^{\infty}$. It may be that the overlapping bridge of V_n is shorter than the nonoverlapping bridge. It may even be that the thickness determined by V_n in the ordering $\{V_n\}_{n=1}^{\infty}$ is less than τ.

Proof. We will prove this lemma only for the even gaps $\{V_{2n}\}_{n=1}^{\infty}$. The proofs for the other three subsequences are similar.

Since $\{V_{2n}\}_{n=1}^{\infty}$ is ordered by decreasing size, Lemma 2.5 implies that when the gaps $\{V_{2n}\}_{n=1}^{\infty}$ are removed in that order, each gap V_{2n} determines a thickness greater than or equal to τ. So in particular, if we let R_{2n} denote the bridge of V_{2n}, in the ordering $\{V_{2n}\}_{n=1}^{\infty}$, that does not contain x, then

$$\frac{|R_{2n}|}{|V_{2n}|} \geq \tau.$$

But the bridge of V_{2n} in the ordering $\{V_{2n}\}_{n=1}^{\infty}$ that does not contain x is the same interval as the nonoverlapping bridge that the gap V_{2n} has in the ordering $\{V_i\}_{i=1}^{\infty}$ (when $i = 2n$). See, for example, the following picture.

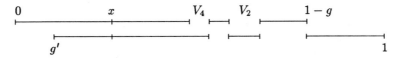

$\square$

Now we shall explain why this last lemma is useful. What we would like to be able to do is compare the sequence of points (h_n, h_n') with the sequence $(g_n, g_n') = \Psi_{\tau\tau'}^n(g, g')$, but we can not do so because the sequences of gaps $\{V_n\}_{n=1}^{\infty}$ and $\{V_n'\}_{n=1}^{\infty}$ are not ordered by decreasing size. However, the last lemma does allow us to compare (h_n, h_n') with the sequence $\overline{\Psi}_{\tau\tau'}^n(g, g')$ (where $\overline{\Psi}_{\tau\tau'}$ is the map defined in Section 7, just before Lemma 7.4). The reason for this is that the last lemma tells us about the ratios of the gaps V_n and V_n' with their nonoverlapping bridges, and the map $\overline{\Psi}_{\tau\tau'}$ comes from a modification of the geometric process so that gaps U_n and U_n' defined by the modified geometric process always determine ratios with their nonoverlapping bridges exactly equal to τ and τ' respectively (in other words, the modified geometric process is only concerned with the nonoverlapping bridges and it ignores the overlapping ones).

Lemma 9.7. *Suppose that Γ and Γ' are Cantor sets that satisfy assumptions (1)–(4). Then for any $n \geq 0$ we have*

$$(h_{n+1}, h_{n+1}') \leq \overline{\Psi}_{\tau\tau'}(h_n, h_n')$$

where we are again using the order relation $(x_1, y_1) \leq (x_2, y_2)$ if and only if $x_1 \leq x_2$ and $y_1 \leq y_2$.

Proof. Consider the following two pictures.

The one on the left represents the gaps V_{n+1} and V_{n+1}' in the overlapping bridges A_n and A_n' of the given Cantor sets Γ and Γ'. The picture on the right represents the modified geometric process mentioned just before this lemma, applied to the intervals A_n and A_n'. By Lemma 6 of this section, the ratios of the lengths of V_{n+1} and V_{n+1}' with their nonoverlapping bridges is greater than or equal to τ and τ' respectively. By the definition of the modified geometric process, the ratios of the lengths of U_{n+1} and U_{n+1}' with their nonoverlapping bridges is exactly τ and τ' respectively. Since the intervals these gaps are removed from are the same in both pictures, we can conclude that $\alpha_{n+1} \le |U_{n+1}|$ and that $\alpha'_{n+1} \le |U'_{n+1}|$ (it need not be true that $|V_{n+1}| \le |U_{n+1}|$ or $|V'_{n+1}| \le |U'_{n+1}|$). Then, since $|A_{n+1} \cup A'_{n+1}| = |B_{n+1} \cup B'_{n+1}|$, when we divide by this length to determine h_{n+1}, h'_{n+1}, and $\overline{\Psi}_{\tau\tau'}(h_n, h'_n)$, we see that $h_{n+1} \le \pi_1(\overline{\Psi}_{\tau\tau'}(h_n, h'_n))$ and that $h'_{n+1} \le \pi_2(\overline{\Psi}_{\tau\tau'}(h_n, h'_n))$. $\qquad\square$

Lemma 9.8. *Suppose that Γ and Γ' are Cantor sets that satisfy assumptions (1)–(4). Then for any $N \ge 0$ and for all $n \ge N$,*

$$(h_n, h'_n) \le \overline{\Psi}_{\tau\tau'}^{\,n-N}(h_N, h'_N).$$

Proof. Choose any $N \ge 0$ and then choose any $n \ge N$. Using the fact that the map $\overline{\Psi}_{\tau\tau'}$ preserves the order relation, and by repeated use of the last lemma, we get

$$
\begin{aligned}
(h_n, h'_n) &\le \overline{\Psi}_{\tau\tau'}(h_{n-1}, h'_{n-1}) \\
&\le \overline{\Psi}_{\tau\tau'}\left(\overline{\Psi}_{\tau\tau'}(h_{n-2}, h'_{n-2})\right) \\
&= \overline{\Psi}_{\tau\tau'}^{\,2}(h_{n-2}, h'_{n-2}) \\
&\vdots \\
&\le \overline{\Psi}_{\tau\tau'}^{\,n-N}(h_{n-(n-N)}, h'_{n-(n-N)}). \qquad\square
\end{aligned}
$$

Lemma 9.9. *Suppose that Γ and Γ' are Cantor sets that satisfy assumptions (1)–(4) and suppose that there is an N such that $\{V_n\}_{n=1}^N$ and $\{V'_n\}_{n=1}^N$ are ordered by decreasing size. Then*

$$(h_n, h'_n) \le \Psi_{\tau\tau'}^n(g, g') \qquad \text{for } 0 \le n \le N$$

and

$$(h_n, h'_n) \le \overline{\Psi}_{\tau\tau'}^{\,n-N}\left(\Psi_{\tau\tau'}^N(g, g')\right) \qquad \text{for } n > N.$$

Proof. The proof is by induction. For $n = 0$ we have

$$(h_0, h'_0) = (g, g') = \Psi_{\tau\tau'}^0(g, g').$$

Now suppose that for some $0 \le n < N$ we have

$$(h_n, h'_n) \le \Psi_{\tau\tau'}^n(g, g').$$

We need to show that

$$(h_{n+1}, h'_{n+1}) \le \Psi_{\tau\tau'}^{n+1}(g, g').$$

It will suffice to prove the following claim.

Claim. $(h_{n+1}, h'_{n+1}) \le \Psi_{\tau\tau'}(h_n, h'_n)$.

For if we combine the claim with the induction step and the fact that $\Psi_{\tau\tau'}$ preserves the order relation (Lemma 6.9), we get

$$(h_{n+1}, h'_{n+1}) \leq \Psi_{\tau\tau'}(h_n, h'_n) \leq \Psi_{\tau\tau'}(\Psi^n_{\tau\tau'}(g, g')).$$

To prove the claim, consider the following two pictures.

The one on the left represents removing gaps V_{n+1} and V'_{n+1} from the bridges A_n and A'_n. The picture on the right represents applying the geometric process from Section 6 (with thicknesses τ and τ') to the intervals A_n and A'_n. By our hypothesis that $\{V_n\}^N_{n=1}$ and $\{V'_n\}^N_{n=1}$ are ordered by decreasing size, we know that the gaps V_{n+1} and V'_{n+1} determine thicknesses greater than or equal to τ and τ'. And by the definition of the geometric process, gaps U_{n+1} and U'_{n+1} determine thicknesses exactly equal to τ and τ'. Since the fundamental intervals (i.e., A_n and A'_n) are the same in both pictures, we can conclude that $\alpha_{n+1} \leq |U_{n+1}|$ and that $\alpha'_{n+1} \leq |U'_{n+1}|$. When we divide everything by $|A_{n+1} \cup A'_{n+1}| = |B_{n+1} \cup B'_{n+1}|$ we get $(h_{n+1}, h'_{n+1}) \leq \Psi_{\tau\tau'}(h_n, h'_n)$ which proves the claim.

The second conclusion of this lemma follows from the first conclusion of this lemma, the previous lemma and the fact that $\overline{\Psi}_{\tau\tau'}$ preserves the order relation. $\square$

Corollary 9.10. *Suppose that Γ and Γ' are Cantor sets that satisfy assumptions (1)–(4). Then for $n \geq 1$*

$$(h_n, h'_n) \leq \overline{\Psi}^{n-1}_{\tau\tau'}\left(\Psi_{\tau\tau'}(g, g')\right).$$

Proof. Clearly, the (very short) sequences $\{V_n\}^1_{n=1}$ and $\{V_n\}^1_{n=1}$ are ordered by decreasing size. So we can apply the previous lemma with $N = 1$ to get the result. $\square$

Lemma 9.11. *Suppose that Γ and Γ' are Cantor sets that satisfy assumptions (1)–(4). Then for all n the point (h_n, h'_n) must be on or above the line of period two points of $\overline{\Psi}_{\tau\tau'}$.*

Proof. Suppose that for some N the point (h_N, h'_N) is strictly below the line of period two points of $\overline{\Psi}_{\tau\tau'}$. Then by Lemma 7.8, the point (h_N, h'_N) is in the basin of attraction of the hyperbolic sink of $\overline{\Psi}_{\tau\tau'}$ at the origin. This implies that the infinite series

$$\sum^{\infty}_{n=N} \left[\pi_1\left(\overline{\Psi}^{n-N}_{\tau\tau'}(h_N, h'_N)\right) + \pi_2\left(\overline{\Psi}^{n-N}_{\tau\tau'}(h_N, h'_N)\right) \right]$$

converges. But we also know from Lemma 8 that for $n \geq N$

$$(h_n, h'_n) \leq \overline{\Psi}^{n-N}_{\tau\tau'}(h_N, h'_N).$$

So the infinite series

$$\sum^{\infty}_{n=N} (h_n + h'_n)$$

also converges. So then the infinite product

$$\prod_{n=0}^{\infty} [1 - (h_n + h'_n)]$$

must converge to a nonzero number. But this implies, by Lemma 9.2, that the lengths $|A_n \cup A'_n|$ do not go to zero as $n \to \infty$. So there is a closed interval J such that J is contained in $A_n \cup A'_n$ for every n and such that $x \in J$. Now we will show that there is no gap of Γ contained in the interval J, which contradicts that Γ is a Cantor set.

Suppose there is a gap W of Γ contained in J. The gap W must be on one side of the point x. We will assume that W is on the right hand side of x. Since the lengths of the gaps V_n must go to zero as $n \to \infty$, we can find an N such that $|V_{2N}| < |W|$ and $|V_{2N-1}| < |W|$. Notice that V_{2N} is on the same side of x as W. So we have the following picture of the bridges A_{2N-2} and A'_{2N-2}.

Lemma 3 from this section, applied to A_{2N-2} and A'_{2N-2}, implies that either V_{2N-1} or V_{2N} is the largest gap of Γ to intersect the interval A'_{2N-2}. But this contradicts that $|V_{2N}| < |W|$ and $|V_{2N-1}| < |W|$. So there cannot be a gap of Γ contained in J. □

In all the lemmas we have proven so far in this section, we have assumed that the Cantor sets Γ and Γ' satisfy the assumptions (1)–(4). The next lemma will show that these assumptions place restrictions on the pairs (τ, τ') and (g, g').

Lemma 9.12. *Suppose Γ and Γ' are Cantor sets that satisfy assumptions (1)–(4). Then (τ, τ') and (g, g') cannot be such that the point $\Psi_{\tau\tau'}(g, g')$ is below the line of period two points of the map $\overline{\Psi}_{\tau\tau'}$.*

So in particular, τ and τ' cannot be such that the image of the map $\Psi_{\tau\tau'}$ is below the line of period two points of $\overline{\Psi}_{\tau\tau'}$.

Proof. Suppose that τ and τ' are such that the point $\Psi_{\tau\tau'}(g, g')$ is below the line of period two points of $\overline{\Psi}_{\tau\tau'}$. Then by Corollary 10 we know that for $n \geq 1$, the points (h_n, h'_n) are all below the line of period two points. But this contradicts the last lemma. So in particular, part of the image of $\Psi_{\tau\tau'}$ must be on or above the line of period two points. □

Now we shall determine for which parameter values τ and τ' it is true that the image of $\Psi_{\tau\tau'}$ is below the line of period two points of $\overline{\Psi}_{\tau\tau'}$.

Lemma 9.13. *The set of parameter values τ and τ' for which the image of $\Psi_{\tau\tau'}$ is below the line of period two points of $\overline{\Psi}_{\tau\tau'}$ is given by*

$$\Lambda = \left\{ (\tau, \tau') \mid 0 < \tau^2 \tau'^2 - 4\tau\tau' - 2\tau - 2\tau' - 1 \right\}.$$

Proof. The line of period two points of $\overline{\Psi}_{\tau\tau'}$ has the following equation (see Lemma 7.6).

$$g' = \frac{\tau\tau' - 1}{\tau'(1+\tau)} - g\frac{\tau(1+\tau')}{\tau'(1+\tau)}$$

The upper right hand corner of the image of $\Psi_{\tau\tau'}$ has the coordinates $(\,1/(1+\tau), 1/(1+\tau')\,)$. So we need to solve the following inequality

$$\frac{1}{1+\tau'} < \frac{\tau\tau' - 1}{\tau'(1+\tau)} - \left(\frac{1}{1+\tau}\right)\frac{\tau(1+\tau')}{\tau'(1+\tau)}$$

which determines the set of parameter values for which the image of $\Psi_{\tau\tau'}$ is below the line of period two points of $\overline{\Psi}_{\tau\tau'}$. But this inequality is equivalent to the one in the definition of Λ.

Notice that the inequality is symmetric in τ and τ', so the set Λ is symmetric about the line $\tau = \tau'$. □

The next thing we want to do is determine the relationship between the set Λ and the sets Λ_1 and Λ_2. If we rewrite the inequality in the definition of Λ as

$$0 < (\tau^2)\tau'^2 - 2(1+2\tau)\tau' - (1+2\tau)$$

then we can use the quadratic formula to solve for τ' as a function of τ. Then we get the inequality

$$\tau' > \frac{1+2\tau}{\tau^2} + \frac{(1+\tau)\sqrt{1+2\tau}}{\tau^2}. \tag{$*$}$$

Recall that the inequality

$$\tau' \le \frac{\tau^2 + 3\tau + 1}{\tau^2}$$

defines part of the boundary between the sets Λ_1 and Λ_2 when $\tau \le 1 + \sqrt{2}$. If we rewrite this inequality as

$$\tau' \le \frac{1+2\tau}{\tau^2} + \frac{(1+\tau)\sqrt{\tau^2}}{\tau^2}$$

then for values of τ less than or equal to $1+\sqrt{2}$, we can easily compare this inequality with the one above labeled ($*$). We see that the equations intersect only when $\tau^2 - 2\tau - 1 = 0$, i.e., when $\tau = 1 + \sqrt{2}$ and, by symmetry, when $\tau' = 1 + \sqrt{2}$ also.

The other part of the boundary between the sets Λ_1 and Λ_2 is defined by the inequality

$$\tau' \le \frac{(1+2\tau)^2}{\tau^3}$$

when $\tau \ge 1 + \sqrt{2}$. To compare this boundary with the boundary of Λ, we need to solve the equation

$$\frac{(1+2\tau)^2}{\tau^3} = \frac{1+2\tau}{\tau^2} + \frac{(1+\tau)\sqrt{1+2\tau}}{\tau^2}$$

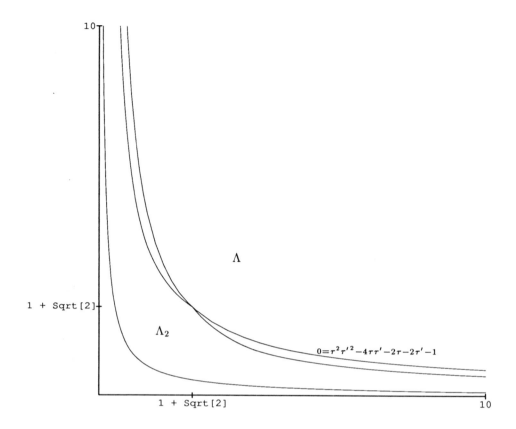

Figure 9

which is equivalent to the polynomial equation

$$0 = \tau^2 - 2\tau - 1.$$

Once again, we see that the only intersection point is when $\tau = 1+\sqrt{2}$ and $\tau' = 1+\sqrt{2}$. We have shown that the set Λ intersects the set Λ_2 only at the point $(\tau, \tau') = (1+\sqrt{2}, 1+\sqrt{2})$. See Figure 9 for the graph of these sets. The upper region is Λ.

The last two lemmas proven above tell us that the conclusion of Theorem 1 from this section is true in the upper region of Figure 9. But the upper region (i.e., Λ) does not fill up the set Λ_1. So now we need to work on proving Theorem 1 for the wedge between Λ and Λ_2.

If we choose (τ, τ') in the wedge between Λ and Λ_2, then part of the image of $\Psi_{\tau\tau'}$ will extend over the line of period two points of $\overline{\Psi}_{\tau\tau'}$. See Figure 10, where $\tau = 5.5$ and $\tau' = 1$ and the dashed line is the line of period two points of $\overline{\Psi}_{\tau\tau'}$. Notice that the image of $\Psi_{\tau\tau'}$

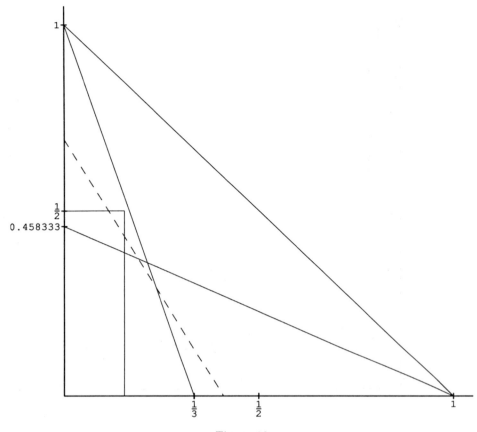

Figure 10

will extend over the line of period two points of $\overline{\Psi}_{\tau\tau'}$ if and only if region (i) also extends over the line of period two points. By Lemma 12, we know that the conclusion of Theorem 1 is true for those points (g, g') in the domain of $\Psi_{\tau\tau'}$ which have their image $\Psi_{\tau\tau'}(g, g')$ below the line of period two points of $\overline{\Psi}_{\tau\tau'}$. So to prove Theorem 1 for (τ, τ') in the wedge between Λ and Λ_2, we only need to worry about those points (g, g') in the domain of $\Psi_{\tau\tau'}$ which have their image $\Psi_{\tau\tau'}(g, g')$ on or above the line of period two points of $\overline{\Psi}_{\tau\tau'}$.

In order to make the next few proofs easier, we will define two new sequences of gaps, called $\{W_n\}_{n=1}^{\infty}$ and $\{W_n'\}_{n=1}^{\infty}$, that will replace the sequences $\{V_n\}_{n=1}^{\infty}$ and $\{V_n'\}_{n=1}^{\infty}$. Given the two Cantor sets Γ and Γ' that satisfy assumptions (1)–(4), define W_1 and W_1' to be gaps that have lengths α_1 and α_1' and such that the right hand end points of W_1 and V_1 are the same and the left hand end points of W_1' and V_1' are also the same. Then the left hand endpoint of W_1 is the point g' and the right hand endpoint of W_1' is the point $1 - g$. See the following two pictures for a comparison of these gaps.

$$0 \quad V_1 \qquad\qquad 1-g \qquad\qquad 0 \quad W_1 \qquad\qquad 1-g$$

$$g' \qquad\qquad V_1' \quad 1 \qquad\qquad g' \qquad\qquad W_1' \quad 1$$

Now do the same thing at the next level of the Cantor sets, replacing gaps V_2 and V_2' with new gaps W_2 and W_2'. Repeat this process with all of the gaps V_n and V_n'. Notice that this process does not change the bridges A_n and A_n' nor does it effect the numbers h_n and h_n'. Also, since we have always made the new bridges determined by the W_n and W_n' at least as long as the bridges determined by the V_n and V_n', if we order the new gaps W_n and W_n' by decreasing size and take the infimum of all the defined thicknesses, we will still get numbers greater than or equal to τ and τ'.

Lemma 9.14. *Suppose that Cantor sets Γ and Γ' satisfy assumptions (1)–(4). Suppose that for some $n \geq 1$ we have*

$$h_n \geq 1 - h_n'(2 + \tau') \qquad \text{and} \qquad h_n' \geq 1 - h_n(2 + \tau).$$

Then $|W_{n+1}| \leq |W_n|$ and $|W_{n+1}'| \leq |W_n'|$.

Proof. Suppose that $|W_{n+1}| > |W_n|$. We are going to consider what happens when we remove just the gaps W_n and W_{n+1} from the fundamental interval $[0, 1 - g]$. Since we are assuming that W_{n+1} is longer than W_n, we should remove W_{n+1} first and then remove W_n. Lemma 2.5 tells us that in this order, both gaps will determine a thickness greater than or equal to τ. The bridge that is between W_{n+1} and W_n is the bridge we are calling A_{n+1}. So when we remove the gap W_n after removing W_{n+1} we should have

$$\frac{|A_{n+1}|}{|W_n|} \geq \tau.$$

From the following picture

$$W_{n+1} \qquad A_{n+1} \qquad W_n$$

$$W_n' \qquad A_{n+1}' \qquad W_{n+1}'$$

we see that

$$\begin{aligned}
\frac{|A_{n+1}|}{|W_n|} &= \frac{|A_n \cup A_n'| - |W_n| - |W_{n+1}| - |W_n'|}{|W_n|} \\
&< \frac{|A_n \cup A_n'| - |W_n| - |W_n| - |W_n'|}{|W_n|} \\
&= \frac{|A_n \cup A_n'| - 2\alpha_n - \alpha_n'}{\alpha_n} \\
&= \frac{1 - 2h_n - h_n'}{h_n}
\end{aligned}$$

$$\leq \frac{1 - 2h_n - [1 - h_n(2 + \tau)]}{h_n}$$

$$= \tau.$$

But now we have the contradiction

$$\frac{|A_{n+1}|}{|W_n|} \geq \tau \quad \text{and} \quad \frac{|A_{n+1}|}{|W_n|} < \tau.$$

So we must have $|W_{n+1}| \leq |W_n|$. The proof that $|W'_{n+1}| \leq |W'_n|$ is similar.   □

This last lemma implies that, if for some $N \geq 1$ the points $\{(h_n, h'_n)\}_{n=1}^N$ stay in the part of the domain of $\Psi_{\tau\tau'}$ that is defined by the inequalities

$$g \geq 1 - g'(2 + \tau') \quad \text{and} \quad g' \geq 1 - g(2 + \tau), \qquad (**)$$

then the first N gaps $\{W_n\}_{n=1}^N$ and $\{W'_n\}_{n=1}^N$ will be ordered by decreasing size.

Suppose that (τ, τ') is in the wedge between Λ and Λ_2. We know that (g, g') must be such that $\Psi_{\tau\tau'}(g, g')$ is on or above the line of period two points of $\overline{\Psi}_{\tau\tau'}$. We also know that the origin is the only fixed point of $\Psi_{\tau\tau'}$ (since $(\tau, \tau') \in \Lambda_1$). So by Theorem 7.1(4) there is an N such that $\Psi_{\tau\tau'}^N(g, g')$ is below the line of period two points of $\overline{\Psi}_{\tau\tau'}$. If we can show that the points $\{(h_n, h'_n)\}_{n=1}^{N-1}$ stay in the part of the domain of $\Psi_{\tau\tau'}$ that is defined by the inequalities $(**)$, then we will know that $\{W_n\}_{n=1}^N$ and $\{W'_n\}_{n=1}^N$ are ordered by decreasing size, and then Lemma 9 (applied with gaps $\{W_n\}_{n=1}^\infty$ and $\{W'_n\}_{n=1}^\infty$) will tell us that (h_N, h'_N) is below the line of period two points of $\overline{\Psi}_{\tau\tau'}$. But this would contradict Lemma 11, so we will have shown that (τ, τ') cannot be in the wedge between Λ and Λ_2.

The next thing we shall do is to determine conditions on τ and τ' such that when part of the image of $\Psi_{\tau\tau'}$ is over the line of period two points of $\overline{\Psi}_{\tau\tau'}$, then that part of the image will also satisfy the inequalities $(**)$. The region in the domain of $\Psi_{\tau\tau'}$ determined by the inequalities $(**)$ is shown in Figure 11, where the dashed lines are the lines determined by the inequalities. The part of the image of $\Psi_{\tau\tau'}$ that is over the line of period two points of $\overline{\Psi}_{\tau\tau'}$ is always a triangle and the three sides are (see Figure 10); a piece of the line of period two points, a piece of the vertical line $g = 1/(1 + \tau)$ and a piece of the horizontal line $g' = 1/(1 + \tau')$. To prove that this triangle is in the region determined by the inequalities $(**)$, it suffices to show that the two corners of the triangle on the line of period two points satisfy the inequalities.

The horizontal line $g' = 1/(1 + \tau')$ and the line of period two points intersect when

$$\frac{1}{1 + \tau'} = \frac{\tau\tau' - 1}{\tau'(1 + \tau)} - g\frac{\tau(1 + \tau')}{\tau'(1 + \tau)},$$

that is, when

$$g = \frac{\tau\tau'^2 - 2\tau' - 1}{\tau(1 + \tau')^2}.$$

Now we need to determine for which τ and τ' the corner of the triangle that this point represents satisfies the inequality $g' \geq 1 - g(2 + \tau)$. So we need to solve

$$\frac{1}{1 + \tau'} \geq 1 - \left(\frac{\tau\tau'^2 - 2\tau' - 1}{\tau(1 + \tau')^2}\right)(2 + \tau).$$

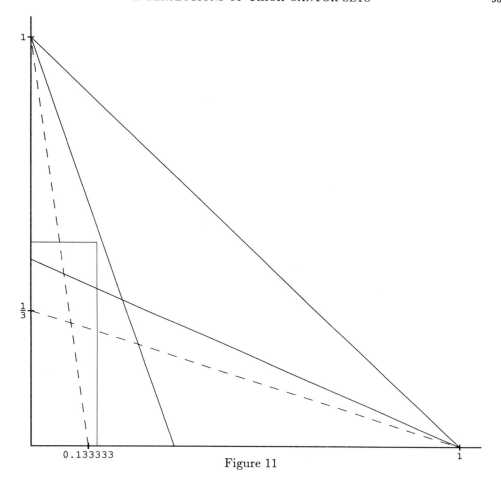

Figure 11

However, it is a little bit easier to notice that the line $g' = 1 - g(2 + \tau)$ intersects the horizontal line $g' = 1/(1 + \tau')$ at the point

$$g = \frac{\tau'}{(1 + \tau')(2 + \tau)}$$

and then solve the inequality

$$\frac{\tau'}{(1 + \tau')(2 + \tau)} \leq \frac{\tau \tau'^2 - 2\tau' - 1}{\tau(1 + \tau')^2}$$

which is equivalent to

$$0 \leq \tau^2 \tau'^2 + \tau \tau'^2 - 3\tau\tau' - 4\tau' - \tau - 2.$$

In a similar way, when the image of $\Psi_{\tau\tau'}$ is partially over the line of period two points the other corner of the triangle will satisfy the inequality $g \geq 1 - g'(2 + \tau')$ when the parameter values τ and τ' satisfy the inequality

$$0 \leq \tau^2 \tau'^2 + \tau^2 \tau' - 3\tau\tau' - 4\tau - \tau' - 2.$$

Notice that when we let $\tau = \tau'$, the equations

$$0 = \tau^2 \tau'^2 + \tau\tau'^2 - 3\tau\tau' - 4\tau' - \tau - 2 \tag{1}$$

and

$$0 = \tau^2 \tau'^2 + \tau^2 \tau' - 3\tau\tau' - 4\tau - \tau' - 2 \tag{2}$$

both become

$$0 = \tau^4 + \tau^3 - 3\tau^2 - 5\tau - 2$$

which factors into

$$0 = (\tau - 2)(\tau + 1)^3$$

which shows that equations (1) and (2) both intersect the line $\tau = \tau'$ at the point $(2, 2)$. On the other hand, if we have a point (τ, τ') that satisfies both of these equations, then

$$\tau^2 \tau'^2 + \tau\tau'^2 - 3\tau\tau' - 4\tau' - \tau - 2 = \tau^2 \tau'^2 + \tau^2 \tau' - 3\tau\tau' - 4\tau - \tau' - 2$$

$$\tau\tau'^2 - 4\tau' - \tau = \tau^2 \tau' - 4\tau - \tau'$$

$$\tau\tau'^2 - 3\tau' = \tau^2 \tau' - 3\tau$$

$$\tau'(\tau\tau' - 3) = \tau(\tau\tau' - 3)$$

$$\tau' = \tau.$$

So the only point that satisfies both equations (1) and (2) is $(2, 2)$ (it's not hard to show that the graph of $\tau\tau' = 3$ never intersects the graphs of equations (1) or (2)). This also shows that when $\tau' \leq \tau$, the graph of equation (1) is above the graph of equation (2) (because if (τ, τ') satisfies both $\tau' \leq \tau$ and equation (1), then (τ, τ') also satisfies the inequality $0 \leq \tau^2 \tau'^2 + \tau^2 \tau' - 3\tau\tau' - 4\tau - \tau' - 2$, which shows that the graph of equation (1) is above the graph of equation (2)).

Lemma 9.15. *Suppose that (τ, τ') is in the wedge between Λ and Λ_2. Then (τ, τ') satisfies the two inequalities*

$$0 \leq \tau^2 \tau'^2 + \tau\tau'^2 - 3\tau\tau' - 4\tau' - \tau - 2$$

and

$$0 \leq \tau^2 \tau'^2 + \tau^2 \tau' - 3\tau\tau' - 4\tau - \tau' - 2.$$

In other words, the triangular part of the image of $\Psi_{\tau\tau'}$ that is on or above the line of period two points of $\overline{\Psi}_{\tau\tau'}$ is still contained in the region defined by the two inequalities

$$g \geq 1 - g'(2 + \tau') \qquad and \qquad g' \geq 1 - g(2 + \tau). \tag{**}$$

Proof. We shall prove this lemma for parameter values τ and τ' that satisfy $\tau \geq \tau'$, i.e., for just one piece of the wedge.

When $\tau \geq \tau'$, one part of the boundary of Λ_2 is defined by the equation

$$\tau = \frac{\tau'^2 + 3\tau' + 1}{\tau'^2}.$$

We will show that the part of the curve defined by equation (1)

$$0 = \tau^2 \tau'^2 + \tau \tau'^2 - 3\tau\tau' - 4\tau' - \tau - 2 \tag{1}$$

that is in the set $\tau \geq \tau'$ is always below the curve

$$\tau = \frac{\tau'^2 + 3\tau' + 1}{\tau'^2}.$$

Rewrite equation (1) as

$$0 = (\tau'^2)\tau^2 + (\tau'^2 - 3\tau' - 1)\tau - (2 + 4\tau')$$

and then use the quadratic formula to solve for τ as a function of τ'. So

$$\tau = \frac{(1 + 3\tau' - \tau'^2) + \sqrt{(1 + 3\tau' - \tau'^2)^2 + 8(1 + 2\tau')\tau'^2}}{2\tau'^2}$$

$$= \frac{(1 + 3\tau' - \tau'^2) + \sqrt{(1 + 3\tau')^2 + (\tau'^4 + 10\tau'^3 + 6\tau'^2)}}{2\tau'^2}.$$

Now we need to show that

$$\frac{(1 + 3\tau' - \tau'^2) + \sqrt{(1 + 3\tau')^2 + (\tau'^4 + 10\tau'^3 + 6\tau'^2)}}{2\tau'^2} \leq \frac{\tau'^2 + 3\tau' + 1}{\tau'^2}$$

or

$$(1 + 3\tau' - \tau'^2) + \sqrt{(1 + 3\tau')^2 + (\tau'^4 + 10\tau'^3 + 6\tau'^2)} \leq 2(\tau'^2 + 3\tau' + 1).$$

This simplifies to $0 \leq 8\tau'^3(\tau' + 1)$ which is true for all positive τ'. $\qquad\square$

Now we can prove the contrapositive of Theorem 1.

Lemma 9.16. *Suppose that Γ and Γ' are Cantor sets that satisfy assumptions (1)–(4). Then (τ, τ') is not in Λ_1.*

Proof. By Lemmas 12 and 13, (τ, τ') is not in Λ. So assume that (τ, τ') is in the wedge between Λ and Λ_2. By Lemma 12, we know that (g, g') cannot be such that $\Psi_{\tau\tau'}(g, g')$ is below the line of period two points of $\overline{\Psi}_{\tau\tau'}$. So assume that $\Psi_{\tau\tau'}(g, g')$ is above the line of period two points of $\overline{\Psi}_{\tau\tau'}$.

The image of $\Psi_{\tau\tau'}$ is partitioned into two subsets, the part that is strictly below the line of period two points and the triangular part that is on or above the line of period two points. By the last lemma, the triangular piece of the image of $\Psi_{\tau\tau'}$ is contained in the part of the domain of $\Psi_{\tau\tau'}$ that is defined by the inequalities $(**)$.

By Corollary 10, we have $(h_1, h_1') \leq \Psi_{\tau\tau'}(g, g')$. By Lemma 11, the point (h_1, h_1') cannot be below the line of period two points so it must be in the triangular piece of the image of $\Psi_{\tau\tau'}$. Then by Lemma 14, the sequences of gaps $\{W_n\}_{n=1}^2$ and $\{W_n'\}_{n=1}^2$ are ordered by decreasing size.

Then by Lemma 9, applied to the gaps $\{W_n\}_{n=1}^2$ and $\{W_n'\}_{n=1}^2$, we have $(h_2, h_2') \leq \Psi_{\tau\tau'}^2(g, g')$. Since the point (h_2, h_2') cannot be below the line of period two points, it must be in the triangular piece of the image of $\Psi_{\tau\tau'}$. Then by Lemma 14, the sequences of gaps $\{W_n\}_{n=1}^3$ and $\{W_n'\}_{n=1}^3$ are ordered by decreasing size.

Continuing in this way using induction, we see that

$$(h_n, h_n') \leq \Psi_{\tau\tau'}^n(g, g')$$

for all $n \geq 1$. But, since (τ, τ') is in Λ_1, by Theorem 7.1(4) there is an N such that $\Psi_{\tau\tau'}^N(g, g')$ is below the line of period two points. This implies that (h_N, h_N') is below the line of period two points, which contradicts Lemma 11. So (τ, τ') is not in Λ_1.   □

10. The Boundary Between Λ_1 and Λ_2

In this section we will prove parts (3) and (4) of Theorem 6.1, the parts that deal with (τ, τ') on the boundary between Λ_1 and Λ_2. First we will prove part (4) of Theorem 6.1.

Theorem 10.1. *Suppose that Γ and Γ' are Cantor sets with thicknesses τ and τ' and fundamental intervals $I = [0, 1 - g]$ and $I' = [g', 1]$ with $g + g' < 1$. Also suppose that $\Gamma \cap \Gamma'$ contains an overlapped point x of the first kind. If τ and τ' both equal $1 + \sqrt{2}$, then the intersection of Γ and Γ' is an infinite set.*

Proof. Suppose that $\Gamma \cap \Gamma'$ is a finite set. Then there is a δ such that $[x - \delta, x + \delta] \cap (\Gamma \cap \Gamma')$ contains only x. Let $\mathcal{U}$ and $\mathcal{U}'$ be any orderings of the gaps of Γ and Γ' by decreasing size. Choose n large enough so that the bridges of $C_n(\mathcal{U})$ and $C_n(\mathcal{U}')$ that contain x both have length less than δ. Let B and B' denote these bridges. Then the Cantor sets $B \cap \Gamma$ and $B' \cap \Gamma'$ intersect only at x. The thickness of each of these two Cantor sets is still $1 + \sqrt{2}$ because if one of them were greater than $1 + \sqrt{2}$, then the new thicknesses would form a point in Λ_1, which would contradict Theorem 1 from the last section. So we can assume, without lose of generality, that $\Gamma \cap \Gamma'$ is the single point x.

Now let's look at the geometry and dynamics of the map $\Psi_{\tau\tau'}$ when $\tau = \tau' = 1 + \sqrt{2}$. Since $\tau = \tau'$, the regions (i)–(iv) in the domain of $\Psi_{\tau\tau'}$ are symmetric with respect to the line $g = g'$. The image of $\Psi_{\tau\tau'}$ is a square with its upper right hand corner on the line $g = g'$. Since $\tau = \tau' = 1 + \sqrt{2}$, the upper right hand corner of the image of $\Psi_{\tau\tau'}$ is equal to the point where the four regions (i)–(iv) come together. This point is a fixed point and it is the only fixed point of $\Psi_{\tau\tau'}$ other than the origin. The line of period two points of $\overline{\Psi}_{\tau\tau'}$ goes through this fixed point and it has slope -1. The whole image of $\Psi_{\tau\tau'}$ is contained in region (i) and all of region (i) is below the line of period two points of $\overline{\Psi}_{\tau\tau'}$. All the points in region (iv), including it's boundary, are mapped by $\Psi_{\tau\tau'}$ to the nonzero fixed point. All the other points in the domain of $\Psi_{\tau\tau'}$ are in the basin of attraction for the hyperbolic sink at the origin.

The Cantor sets Γ and Γ' satisfy the assumptions (1)–(4) from the last section. So we can apply Corollary 9.10 from the last section to conclude that $(h_1, h_1') \leq \Psi_{\tau\tau'}(g, g')$. Unless $\Psi_{\tau\tau'}(g, g')$ is the nonzero fixed point, the point $\Psi_{\tau\tau'}(g, g')$ will lie below the line of period two points of $\overline{\Psi}_{\tau\tau'}$. But this would contradict Lemma 9.11. So $\Psi_{\tau\tau'}(g, g')$ must equal the fixed point. In order that (h_1, h_1') not lie below the line of period two points of $\overline{\Psi}_{\tau\tau'}$, we must have $(h_1, h_1') = \Psi_{\tau\tau'}(g, g')$. But then the gaps V_1 and V_1' are equal to the gaps U_1 and U_1' defined by the geometric process. This implies that $\Gamma \cap \Gamma'$ contains two nonoverlapped points, which contradicts our assumption that the intersection is a single point. So the intersection cannot be a finite set. $\square$

Theorem 10.2. *Suppose that* Γ *and* Γ' *are Cantor sets with thicknesses* τ *and* τ' *and fundamental intervals* $I = [0, 1 - g]$ *and* $I' = [g', 1]$ *with* $g + g' < 1$. *Also suppose that* $\Gamma \cap \Gamma'$ *contains an overlapped point* x *of the first kind. If* (τ, τ') *is on the piece of the boundary between* Λ_1 *and* Λ_2 *that is defined by the equation*

$$\tau = \frac{\tau'^2 + 3\tau' + 1}{\tau'^2}$$

with $\tau \in (\tau_0, \infty)$, *then the intersection of* Γ *and* Γ' *is an infinite set.*

Proof. Suppose that $\Gamma \cap \Gamma'$ is a finite set. As in the proof of the last theorem, without lose of generality we can assume that $\Gamma \cap \Gamma'$ contains only the point x.

Now let's look at the geometry and dynamics of $\Psi_{\tau\tau'}$ for this choice of (τ, τ'). Since (τ, τ') is on the curve

$$\tau = \frac{\tau'^2 + 3\tau' + 1}{\tau'^2},$$

the upper right hand corner of the image of $\Psi_{\tau\tau'}$ is on the line

$$g = (1 - g')\frac{\tau'}{1 + 2\tau'}$$

(see Lemma 7.3). Since $\tau > \tau'$, the upper right hand corner of the image of $\Psi_{\tau\tau'}$ is on the piece of this line that forms the boundary between regions (iii) and (iv). The upper right hand corner of the image of $\Psi_{\tau\tau'}$ is a fixed point and since $\tau > \tau_0$, it is the only other fixed point besides the origin. All of region (iv), including it's boundary, is mapped by $\Psi_{\tau\tau'}$ to this fixed point. An argument similar to the proof of Lemma 7.12 shows that all the other points in the domain of $\Psi_{\tau\tau'}$ are in the basin of attraction of the hyperbolic sink at the origin. So in particular, unless $\Psi_{\tau\tau'}(g, g')$ is equal to the nonzero fixed point, then (g, g') is in the basin of attraction of the origin.

Part of the image of $\Psi_{\tau\tau'}$ will extend over the line of period two points of $\overline{\Psi}_{\tau\tau'}$ (see Lemma 9.13 from the last section). However, by Lemma 9.15, the part of the image of $\Psi_{\tau\tau'}$ that is over the line of period two points of $\overline{\Psi}_{\tau\tau'}$ is contained in the part of the domain of $\Psi_{\tau\tau'}$ that is defined by the inequalities

$$g \geq 1 - g'(2 + \tau') \qquad \text{and} \qquad g' \geq 1 - g(2 + \tau). \tag{$**$}$$

The Cantor sets Γ and Γ' satisfy the assumptions (1)–(4) from the last section. So by Lemma 9.11, we must have (h_n, h'_n) on or above the line of period two points of $\overline{\Psi}_{\tau\tau'}$ for all n. As long as the points (h_n, h'_n) are above the line of period two points, they satisfy the inequalities $(**)$ which means that, by Lemma 9.14, the sequences of gaps $\{W_n\}_{n=1}^{\infty}$ and $\{W'_n\}_{n=1}^{\infty}$ are ordered by decreasing size. Then, by Lemma 9.9,

$$(h_n, h'_n) \leq \Psi_{\tau\tau'}^n(g, g').$$

This implies that (g, g') must be in region (iv) and that $\Psi_{\tau\tau'}(g, g')$ is the fixed point on the boundary of region (iv) (otherwise, (h_n, h'_n) will eventually be below the line of period two points).

Because the intersection of Γ and Γ' does not contain any nonoverlapped points, we must have strict inequalities

$$(h_n, h'_n) < \Psi^n_{\tau\tau'}(g, g') = \Psi_{\tau\tau'}(g, g'),$$

so in particular $(h_1, h'_1) < \Psi_{\tau\tau'}(g, g')$. This means that (h_1, h'_1) is in the basin of attraction of the origin. If we use the claim from the proof of Lemma 9.9, we get

$$(h_2, h'_2) \leq \Psi_{\tau\tau'}(h_1, h'_1).$$

If we use this claim inductively together with the fact that $\Psi_{\tau\tau'}$ preserves the order relation, we get

$$(h_n, h'_n) \leq \Psi^{n-1}_{\tau\tau'}(h_1, h'_1).$$

Since (h_1, h'_1) is in the basin of attraction of the origin, (h_n, h'_n) must eventually be below the line of period two points of $\overline{\Psi}_{\tau\tau'}$. But this contradicts Lemma 9.11. So the intersection of Γ and Γ' must be an infinite set. $\square$

Theorem 10.3. *Suppose that (τ, τ') is on the graph of the equation*

$$\tau' = \frac{(1 + 2\tau)^2}{\tau^3}$$

for $\tau \in (1 + \sqrt{2}, \tau_0)$. Then there exist Cantor sets Γ and Γ' with thicknesses τ and τ' and fundamental intervals $I = [0, 1 - g]$ and $I' = [g', 1]$, for some $g + g' < 1$, such that $\Gamma \cap \Gamma' = \{x\}$ where x is an overlapped point of the first kind.

Note. The next lemma will deal with the case where $\tau = \tau_0$.

Proof. First let's look at the geometry and dynamics of $\Psi_{\tau\tau'}$ for these parameter values. The upper right hand corner of the image of $\Psi_{\tau\tau'}$ is contained in the interior of region (iii). By Proposition 7.1(3), the map $\Psi_{\tau\tau'}$ has a fixed point with coordinates

$$\left(\frac{\sqrt{\tau\tau'} - 1}{\sqrt{\tau\tau'} + \tau}, \frac{\sqrt{\tau\tau'} - 1}{\sqrt{\tau\tau'} + \tau'} \right).$$

Let q denote this fixed point. This fixed point is on the boundary between regions (i) and (iii) and its first coordinate must equal $1/(1 + \tau)$. It is the only fixed point of $\Psi_{\tau\tau'}$ besides the origin. Now let p be the point where the lines

$$g' = (1 - g)\frac{\tau}{1 + 2\tau} \qquad \text{and} \qquad g = (1 - g')\frac{\tau'}{1 + 2\tau'}$$

intersect, i.e. the point where all four regions in the domain of $\Psi_{\tau\tau'}$ come together. Then the image of p is the upper right hand corner of the image of $\Psi_{\tau\tau'}$ and it lies above the fixed point q on the vertical line $g = 1/(1 + \tau)$. By an argument similar to that used in the proof of Lemma 7.12, we can show that the sequence of points $\Psi^n_{\tau\tau'}(p)$ all lie on the vertical line $g = 1/(1 + \tau)$ above the fixed point q and they converge monotonically downward to q.

The line of period two points of $\overline{\Psi}_{\tau\tau'}$ contains the point q. So all of the image of $\Psi_{\tau\tau'}$ that is contained in region (i) is contained in the basin of attraction of the origin. The basin of attraction of the origin also includes those points in region (iii) whose shadow is to the left of q (the shadow of a point is defined in the proof of Lemma 7.12, right after Claim 4). Those points in region (iii) whose shadow is to the right of q are attracted to the fixed point q and their orbits move strictly downwards on the piece of the vertical line $g = 1/(1 + \tau)$ that is above the point q. So the triangular piece of the image of $\Psi_{\tau\tau'}$ which is above the line connecting the fixed point q with the point $(0, 1)$ is contained in the basin of attraction of q. Let T denote this triangular subset minus the line connecting q to $(0, 1)$.

If we apply the geometric process from Section 6 with the initial point $p = (g_p, g'_p)$ given above, then the geometric process will define two Cantor sets with fundamental intervals $[0, 1 - g_p]$ and $[g'_p, 1]$ and thickness τ and τ' and such that the intersection of these two Cantor sets will contain a unique overlapped point of the first kind and a countable number of nonoverlapped points. To prove the theorem, we need to show that we can shave off the endpoints. However, unlike the previous cases where we have shaved off nonoverlapped points, this time it must be done so that the thicknesses of these two Cantor sets is unchanged.

To do the shaving in this case, we will use the idea of a "sliding" gap that was described in Section 2. We will define a modification of the geometric process that slides gaps a little bit in order to shave off the endpoints. Here is a description of the first iterate of this modified geometric process. Remove the gaps U_1 and U'_1 from the fundamental intervals I and I' as in the geometric process from Section 6. Then slide the gap U_1 over to the left, keeping the thickness constantly equal to τ, by a small amount that will be described later. Then slide the gap U'_1 to the right, keeping the thickness constantly equal to τ', by a small amount that will be described later. Call theses new gaps V_1 and V'_1. These new gaps no longer have g'_p and $1 - g_p$ as endpoints. Notice that the point (h_1, h'_1) defined by the new gaps V_1 and V'_1 is strictly lower and to the left of the point $\Psi_{\tau\tau'}(p)$, i.e.

$$(h_1, h'_1) < \Psi_{\tau\tau'}(p).$$

In the second iterate of the modified geometric process, first apply the geometric process from Section 6 to the bridges R_1 and L'_1 defined by the first iterate of the modified geometric process to get gaps U_2 and U'_2. Then slide these two gaps over by a small amount, that we will describe later, so that the new gaps V_2 and V'_2 do not have common endpoints with the bridges R_1 and L'_1. Then we will also have

$$(h_2, h'_2) < \Psi_{\tau\tau'}(h_1, h'_1).$$

In general, each step of the modified geometric process consists of first applying the geometric process from Section 6 to the overlapping bridges defined by the last step of the modified process and then sliding the gaps away from the overlapping bridges (by an amount that will be described below) to define new gaps V_n and V_n' that do not have endpoints in common with the new overlapping bridges. Then we will always have

$$(h_{n+1}, h_{n+1}') < \Psi_{\tau\tau'}(h_n, h_n').$$

We will now describe how to choose the amounts that the gaps are slid over by in each step of the modified geometric process. The idea is that when we are done, the points (h_n, h_n') defined by the modified geometric process should, in some sense, shadow the orbit under $\Psi_{\tau\tau'}$ of p and always stay in the triangular subset T of the basin of attraction of the fixed point q.

The point $\Psi_{\tau\tau'}(p)$ is the upper right hand corner of the image of $\Psi_{\tau\tau'}$. Slide gaps U_1 and U_1' over so that (h_1, h_1') is somewhere in the interior of T. Then the point $\Psi_{\tau\tau'}(h_1, h_1')$ will be on the vertical line $g = 1/(1 + \tau)$ above the fixed point q. Slide gaps U_2 and U_2' so that (h_2, h_2') is in the interior of the triangle T. In general, the point $\Psi_{\tau\tau'}(h_n, h_n')$ will be on the vertical line $g = 1/(1 + \tau)$ above the fixed point q and the gaps U_{n+1} and U_{n+1}' should be slid over so that the point (h_{n+1}, h_{n+1}') is in the interior of the triangle T. This process can be kept up indefinitely, as long as we never let a point (h_n, h_n') land on the line connecting the points $(0, 1)$ and q. This process defines two interleaved Cantor sets Γ and Γ' with thicknesses τ and τ' and such that their intersection contains a single overlapped point of the first kind. $\square$

Theorem 10.4. *Suppose that*

$$(\tau, \tau') = \left(\tau_0, \frac{(1 + 2\tau_0)^2}{\tau_0{}^3} \right).$$

Then there exist Cantor sets Γ and Γ' with thicknesses τ and τ' and fundamental intervals $I = [0, 1 - g]$ and $I' = [g', 1]$, for some $g + g' < 1$, such that $\Gamma \cap \Gamma' = \{x\}$ where x is an overlapped point of the first kind.

Proof. The proof of this theorem is almost exactly the same as the proof of the last theorem. To see what is different in this case, recall that the value of τ_0 was defined in Section 7 as the first coordinate of the point where the graphs of the equations

$$\tau = \frac{\tau'^2 + 3\tau' + 1}{\tau'^2} \qquad \text{and} \qquad \tau' = \frac{(1 + 2\tau)^2}{\tau^3}$$

intersect off of the line $\tau = \tau'$. When (τ, τ') is this intersection point, the upper right hand corner of the image of $\Psi_{\tau\tau'}$ is on the boundary between regions (iii) and (iv). So the upper right hand corner of the image of $\Psi_{\tau\tau'}$ is a fixed point. This is the main difference between this case and the case in the last theorem.

In this case, instead of choosing the initial point p to be the point where the four regions (i)–(iv) meet (i.e. the point whose image is the upper right hand corner of the image of $\Psi_{\tau\tau'}$), we will choose the initial point p for the modified geometric process to be a point on the vertical line $g = 1/(1 + \tau)$ that is in the interior of region (iii) (that is, strictly below the upper right hand corner of the image of $\Psi_{\tau\tau'}$). For any such choice of p, the modified geometric process described in the proof of the last theorem will complete the proof in this case also. $\square$

11. Proof Of Theorem 1.1

In this section we will prove the four parts of Theorem 1.1.

Lemma 11.1. *Suppose that Γ and Γ' are interleaved Cantor sets with thicknesses τ and τ'. If $(\tau, \tau') \in \Lambda_1$, then there are no isolated overlapped points in $\Gamma \cap \Gamma'$.*

Proof. Suppose that there is an isolated overlapped x in the intersection of Γ and Γ'. We need to consider three cases, depending on which kind of overlapped point x is.

Suppose that x is an overlapped point of the third kind. Let $\mathcal{U} = \{U_i\}_{i=1}^\infty$ and $\mathcal{U}' = \{U_i'\}_{i=1}^\infty$ be orderings of all the gaps of Γ and Γ' by decreasing size. Since x is an isolated overlapped point of the third kind, we can find an n sufficiently large so that the bridges of $C_n(\mathcal{U})$ and $C_n(\mathcal{U}')$ that contain x contain no other overlapped points and so that x is an endpoint of both of these bridges. We will call these bridges B_n and B_n'.

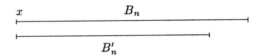

But since Λ_1 is a subset of Λ_3, Theorem 4.1(1) says that x cannot be the only overlapped point in $B_n \cap B_n'$. So x cannot be an overlapped point of the third kind.

Suppose that x is an overlapped point of the second kind. As in the last paragraph, we can find an n sufficiently large so that the bridges B_n and B_n' that contain x do not contain any other overlapped points and so that x is an endpoint of one of the bridges. But since Λ_1 is a subset of Λ_7, Theorem 5.6(1) says that x cannot be the only overlapped point in $B_n \cap B_n'$. So x cannot be an overlapped point of the second kind.

Suppose that x is an overlapped point of the first kind. Again choose n so that the bridges B_n and B_n' that contain x do not contain any other overlapped points. If B_n and B_n' look like the following picture, then Theorem 9.1 gives a contradiction to the assumption that x is the only overlapped point in $B_n \cap B_n'$.

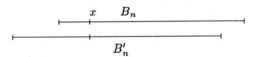

Now suppose that B_n and B_n' look like the following picture.

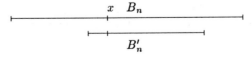

We will show that we can reduce this to the case covered by Theorem 9.1. Let U be the largest gap of $B_n \cap \Gamma$ that intersects B_n'. We will show that U must contain one of the endpoints of B_n'. Suppose that U does not contain an endpoint of the interval B_n'. Then we have the following picture.

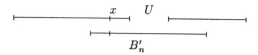

But the left and right hand hand bridges of U in B_n are each clearly interleaved with B_n', which implies by Newhouse's lemma that there are two overlapped points in B_n and B_n', which contradicts our choice of n. So U must contain an endpoint of B_n'. Then Theorem 9.1, applied to the bridge of U that contains x and to the bridge B_n', gives a contradiction to the assumption that x is the only overlapped point in $B_n \cap B_n'$.

So there can be no isolated overlapped points in $\Gamma \cap \Gamma'$.   □

Now we can prove part (1) of Theorem 1.1.

Theorem 11.2. *Suppose that Γ and Γ' are interleaved Cantor sets with thicknesses τ and τ'. If $(\tau, \tau') \in \Lambda_1$, then $\Gamma \cap \Gamma'$ must contain a Cantor set.*

Proof. By the last lemma, there are no isolated overlapped points in $\Gamma \cap \Gamma'$. Then by Lemma 2.8, $\Gamma \cap \Gamma'$ must contain a Cantor set.   □

Now we shall prove part (2) of Theorem 1.1.

Theorem 11.3. *If $(\tau, \tau') \in \Lambda_2$, then there exits interleaved, affine Cantor sets Γ and Γ' with thicknesses τ and τ' such that $\Gamma \cap \Gamma'$ is a single point.*

Proof. In Lemma 8.7 we showed that when (g, g') is a fixed point of $\Psi_{\tau\tau'}$, the geometric process can be used to construct affine Cantor sets with a single overlapped point and a countable number of nonoverlapped points.

Choose $(\tau, \tau') \in \Lambda_2$. Since Λ_2 is open, we can find an ϵ such that $(\tau + \epsilon, \tau' + \epsilon)$ is also in Λ_2. Choose a fixed point (g, g') of $\Psi_{(\tau+\epsilon)(\tau'+\epsilon)}$. We need to show that we can "shave off" the nonoverlapped points from the affine Cantor set given by this fixed point in such a way that the new Cantors sets have thicknesses τ and τ' and are still affine Cantor sets. Let the four gaps U_1, U_1', U_2 and U_2' be defined by the fixed point of $\Psi_{(\tau+\epsilon)(\tau'+\epsilon)}$. Now move the left hand endpoint of U_1 to the left just enough so that the new gap V_1 defines thickness τ. Move the right hand endpoint of U_1' to the right so that the new gap V_1' determines thickness τ'. Then move the right hand endpoint of U_2 to the right so that the new gap V_2 determines thickness τ, and move the left hand endpoint of U_2' to the left so that the new gap V_2' determines thickness τ'. Notice that the bridges L_2 and R_2 are unchanged. Now construct affine Cantor sets Γ and Γ' based on the four gaps V_1, V_1', V_2 and V_2'.   □

Theorem 1.1(3) is proven by applying Theorem 6.1(3), which is proven by Theorems 10.3 and 10.4.

Now we shall prove part (4) of Theorem 1.1. The geometric process from Section 6 gives us the examples we need with countable intersections and only one overlapped point. So we need to prove the statement in Theorem 1.1(4) about the intersection always being infinite.

Theorem 11.4. *If* $(\tau, \tau') = (1 + \sqrt{2}, 1 + \sqrt{2})$ *or if* (τ, τ') *is on the piece of the boundary between* Λ_1 *and* Λ_2 *that is defined by the equation*

$$\tau' = \frac{\tau^2 + 3\tau + 1}{\tau^2}$$

with $\tau \in (\tau_0, \infty)$, *then for any pair of interleaved Cantor sets* Γ *and* Γ' *with thicknesses* τ *and* τ', $\Gamma \cap \Gamma'$ *must contain an infinite number of points.*

Proof. We will prove this theorem by using Theorem 6.1(4), which was proven in Section 10, Theorems 10.1 and 10.2.

Choose (τ, τ') and suppose that Γ and Γ' are interleaved Cantor sets with these thicknesses. If $\Gamma \cap \Gamma'$ does not contain an isolated overlapped point, then the intersection contains a Cantor set so the intersection must be infinite. So assume that the intersection of Γ and Γ' contains an isolated overlapped point x. Since (τ, τ') is in both Λ_3 and Λ_7, x cannot be an overlapped point of the third or second kind (because of Theorems 4.1(2) and 5.6(2) respectively). So x must be an overlapped point of the first kind. But now we can apply Theorem 6.1(4) to conclude that $\Gamma \cap \Gamma'$ is an infinite set. $\square$

Appendix 1

In this appendix we will compute the fixed points of $\Psi_{\tau\tau'}$ that are in region (iii) of the domain of $\Psi_{\tau\tau'}$. Let $\hat{\Psi}_{\tau\tau'}$ denote the function

$$\hat{\Psi}_{\tau\tau'}(g,g') = \left(\frac{1}{1+\tau} , \frac{g}{\tau'(1-g-g')} \right)$$

with domain $\{ (g,g') \mid g,g' \geq 0, g+g' < 1 \}$. The range of $\hat{\Psi}_{\tau\tau'}$ will be the vertical line $g = 1/(1+\tau)$ with $g' \geq 0$. Notice that $\hat{\Psi}_{\tau\tau'}$ is the definition $\Psi_{\tau\tau'}$ in region (iii) extended to the whole domain of $\Psi_{\tau\tau'}$.

To get a fixed point of $\hat{\Psi}_{\tau\tau'}$, we must have $g = 1/(1+\tau)$. To get the second coordinate of a fixed point we need to solve

$$g' = \frac{1/(1+\tau)}{\tau'(1 - 1/(1+\tau) - g')}$$

which is equivalent to the quadratic equation

$$0 = g'^2 - \frac{\tau}{1+\tau}g' + \frac{1}{\tau'(1+\tau)}.$$

Using the quadratic formula to solve for g' as a function of τ and τ', we get

$$g' = \frac{1}{2(1+\tau)} \left(\tau \pm \sqrt{\frac{\tau'\tau^2 - 4\tau - 4}{\tau'}} \right).$$

So $\hat{\Psi}_{\tau\tau'}$ has at most two fixed points. For the fixed points to exist, we must have $\tau'\tau^2 - 4\tau - 4 \geq 0$ or

$$\tau' \geq \frac{4(1+\tau)}{\tau^2}.$$

Notice that for any $k > 0$, $\hat{\Psi}_{\tau\tau'}$ is constant on the lines

$$\frac{g}{1-g-g'} = k$$

or

$$g' = 1 - \left(\frac{1+k}{k} \right) g.$$

ROGER KRAFT

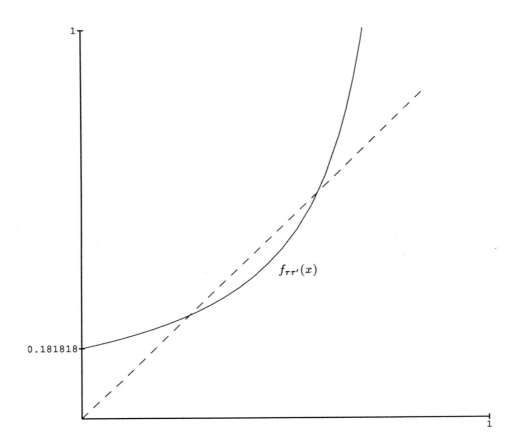

Figure 12

So we see that $\hat{\Psi}_{\tau\tau'}$ is constant on all lines through the point $(0,1)$ with slope strictly less than -1. Each of these lines intersects the vertical line $g = 1/(1+\tau)$ only once, which gives us a semiconjugacy between $\hat{\Psi}_{\tau\tau'}$ and the function

$$f_{\tau\tau'}(x) = \frac{1}{\tau\tau' - x\tau'(1+\tau)}$$

on the domain $\{\, x \mid 0 \le x \le \tau/(1+\tau)\,\}$. The function $f_{\tau\tau'}$ has a saddle-node bifurcation when the parameters (τ, τ') cross the curve

$$\tau' = \frac{4(1+\tau)}{\tau^2}.$$

See Figure 12 for the graph of $f_{\tau\tau'}$ (with $\tau = 5.5$ and $\tau' = 1$).

For the fixed points of $\hat{\Psi}_{\tau\tau'}$ to be fixed points of $\Psi_{\tau\tau'}$, we need to check that they lie in region (iii). We will check each fixed point separately.

First we will check when the upper fixed point of $\hat{\Psi}_{\tau\tau'}$ is in region (iii). The lower boundary of (iii) is defined by the line

$$g' = (1-g)\frac{\tau}{1+2\tau}.$$

For the upper fixed point of $\hat{\Psi}_{\tau\tau'}$ to be above this lower boundary of (iii) we need

$$\frac{1}{2(1+\tau)}\left(\tau + \sqrt{\frac{\tau'\tau^2 - 4\tau - 4}{\tau'}}\right) \geq \left(1 - \frac{1}{1+\tau}\right)\frac{\tau}{1+2\tau}$$

which simplifies to

$$\sqrt{\frac{\tau'\tau^2 - 4\tau - 4}{\tau'}} \geq \frac{-\tau}{1+2\tau}.$$

But this last inequality is always true. So when the fixed points of $\hat{\Psi}_{\tau\tau'}$ exist, the upper one is always above the lower boundary of region (iii).

The upper boundary of region (iii) is defined by the equation

$$g' = 1 - g\frac{1+2\tau'}{\tau'}.$$

For the upper fixed point of $\hat{\Psi}_{\tau\tau'}$ to be below the upper boundary of region (iii), we need

$$\frac{1}{2(1+\tau)}\left(\tau + \sqrt{\frac{\tau'\tau^2 - 4\tau - 4}{\tau'}}\right) \leq 1 - \frac{1}{1+\tau}\frac{1+2\tau'}{\tau'}$$

which simplifies to

$$\sqrt{\frac{\tau'\tau^2 - 4\tau - 4}{\tau'}} \leq \frac{\tau\tau' - 2(1+\tau')}{\tau'}.$$

A necessary condition for this last inequality to be true is that the numerator on the right satisfy $\tau\tau' - 2(1+\tau') \geq 0$ or

$$\tau \geq \frac{2(1+\tau')}{\tau'}.$$

Now if we go back to this last inequality and continue simplifying, we end up with

$$\tau \leq \frac{\tau'^2 + 3\tau' + 1}{\tau'^2}.$$

So for the upper fixed point of $\hat{\Psi}_{\tau\tau'}$ to be in region (iii), we need the parameters τ and τ' to satisfy the two inequalities

$$\tau \geq \frac{2(1+\tau')}{\tau'} \qquad \text{and} \qquad \tau \leq \frac{\tau'^2 + 3\tau' + 1}{\tau'^2}.$$

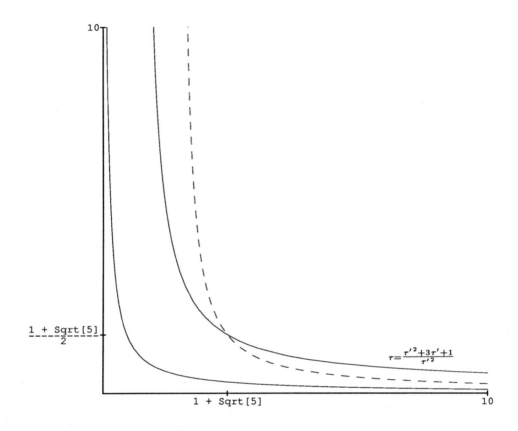

Figure 13

Solving for when these two inequalities intersect reduces to solving $\tau'^2 - \tau' - 1 = 0$ and has the solution $\tau' = (1+\sqrt{5})/2$. The other coordinate of the intersection point is $\tau = 1+\sqrt{5}$. The region in parameter space that satisfies the two inequalities is shown in Figure 13. So the set of parameter values for which the upper fixed point of $\hat{\Psi}_{\tau\tau'}$ is in region (iii) is contained in the set of parameter values for which $\Psi_{\tau\tau'}$ has a fixed point in region (iv).

Now we can check when the lower fixed point of $\hat{\Psi}_{\tau\tau'}$ is in region (iii). For the lower fixed point of $\hat{\Psi}_{\tau\tau'}$ to be above the lower boundary of (iii), we need

$$\frac{1}{2(1+\tau)}\left(\tau - \sqrt{\frac{\tau'\tau^2 - 4\tau - 4}{\tau'}}\right) \geq \left(1 - \frac{1}{1+\tau}\right)\frac{\tau}{1+2\tau}.$$

Similarly to the previous case, this simplifies to

$$-\sqrt{\frac{\tau'\tau^2 - 4\tau - 4}{\tau'}} \geq \frac{-\tau}{1+2\tau}.$$

which can then be further simplified to

$$\tau' \leq \frac{(1 + 2\tau)^2}{\tau^3}.$$

For the lower fixed point to be below the upper boundary of region (iii), we need

$$\frac{1}{2(1 + \tau)} \left(\tau - \sqrt{\frac{\tau'\tau^2 - 4\tau - 4}{\tau'}} \right) \leq 1 - \frac{1}{1 + \tau} \frac{1 + 2\tau'}{\tau'}.$$

This is equivalent to

$$-\sqrt{\frac{\tau'\tau^2 - 4\tau - 4}{\tau'}} \leq \frac{\tau\tau' - 2(1 + \tau')}{\tau'}$$

or

$$\sqrt{\frac{\tau'\tau^2 - 4\tau - 4}{\tau'}} \geq \frac{2(1 + \tau') - \tau\tau'}{\tau'}. \tag{1}$$

We get two cases from this last inequality, depending on the sign of the numerator on the right hand side of (1).

In the first case, when $2(1 + \tau') - \tau\tau' < 0$, inequality (1) will be true. So case (i) is the pair of inequalities

$$\tau' \leq \frac{(1 + 2\tau)^2}{\tau^3} \quad \text{and} \quad \tau > \frac{2(1 + \tau')}{\tau'}$$

or, equivalently

$$\tau' \leq \frac{(1 + 2\tau)^2}{\tau^3} \quad \text{and} \quad \tau' > \frac{2}{\tau - 2}. \tag{i}$$

It can be shown that these two equations intersect only once for $\tau > 0$ and that the intersection is at approximately $\tau = 3.19388$ which is strictly greater than $1 + \sqrt{2}$. The set of parameter values which satisfy these inequalities is shown is Figure 14.

Case (ii) of inequality (1) is when $2(1 + \tau') - \tau\tau' \geq 0$. In this case, we can continue to work on inequality (1), squaring both sides and simplifying it to the inequality

$$\tau \geq \frac{\tau'^2 + 3\tau' + 1}{\tau'^2}.$$

So in case (ii), we have the three inequalities

$$\tau' \leq \frac{(1 + 2\tau)^2}{\tau^3} \quad \text{and} \quad \tau' \leq \frac{2}{\tau - 2} \quad \text{and} \quad \tau \geq \frac{\tau'^2 + 3\tau' + 1}{\tau'^2}. \tag{ii}$$

If we consider just the first and last of these inequalities, we get the subset of the closure of Λ_2 that is between $\tau = 1 + \sqrt{2}$ and $\tau = \tau_0$ and bounded by the two curves

$$\tau' = \frac{(1 + 2\tau)^2}{\tau^3} \quad \text{and} \quad \tau = \frac{\tau'^2 + 3\tau' + 1}{\tau'^2}.$$

See Figure 8 in Section 7.

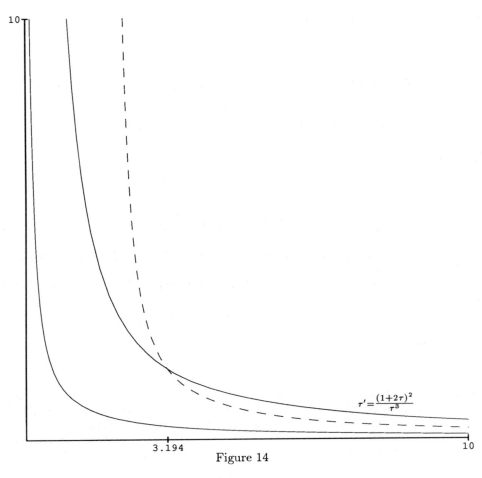

$$\tau' = \frac{(1+2\tau)^2}{\tau^3}$$

Figure 14

So we have shown that in both case (i) and case (ii), the parameter values for which the lower fixed point of $\hat{\Psi}_{\tau\tau'}$ is contained in region (iii) is a subset of the closure of Λ_2.

So all the parameter values for which $\Psi_{\tau\tau'}$ has a fixed point in region (iii) are contained in the closure of Λ_2. We can say the same thing about region (ii) by reversing all the roles of τ and τ' and g and g'.

Appendix 2

This appendix contains an alternate proof of part 2 of Theorem 1.1. The material in this appendix also helps shed some light on the bifurcations of the fixed points of the map $\Psi_{\tau\tau'}$. What we will do is compute (without the use of $\Psi_{\tau\tau'}$) all the possible self-similar constructions of two interleaved Cantor sets that contain only one overlapped point in their intersection (which of course turn out to be all possible fixed points of $\Psi_{\tau\tau'}$). Another way to put it is, we will pick a point (g, g') in the set $\{\, (g,g') \mid g, g' \geq 0,\; g + g' < 1 \,\}$ and then ask for which parameter values τ and τ' is (g, g') a fixed point of $\Psi_{\tau\tau'}$.

We start with the intervals $[0, 1-g]$ and $[g', 1]$ with $g + g' < 1$.

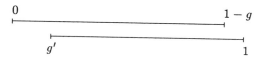

Then we ask, what should be the lengths of the intervals $[1, a]$ and $[-b, 0]$ so that the following picture is "self-similar?"

That is, we want

$$\frac{a}{1+a} = \frac{g'}{1-g} \quad \text{and} \quad \frac{b}{1+b} = \frac{g}{1-g'}.$$

Solving for a and b as functions of g and g', we get

$$a = \frac{g'}{1-g-g'} \quad \text{and} \quad b = \frac{g}{1-g-g'}.$$

The thickness determined by the gap $(1-g, 1)$ in the interval $[0, 1+a]$ is

$$\tau = \begin{cases} \dfrac{a}{g} & \text{if } a \leq 1-g \\[2ex] \dfrac{1-g}{g} & \text{if } a \geq 1-g \end{cases}$$

$$= \begin{cases} \dfrac{g'}{g(1-g-g')} & \text{if } \dfrac{g'}{1-g-g'} \leq 1-g \\[2ex] \dfrac{1-g}{g} & \text{if } \dfrac{g'}{1-g-g'} \geq 1-g \end{cases}$$

113

$$= \begin{cases} \dfrac{g'}{g(1-g-g')} & \text{if } g' \le \dfrac{(1-g)^2}{2-g} \\[3mm] \dfrac{1-g}{g} & \text{if } g' \ge \dfrac{(1-g)^2}{2-g}. \end{cases}$$

Similarly, the thickness determined by the gap $(0, g')$ in the interval $[-b, 1]$ is given by

$$\tau' = \begin{cases} \dfrac{g}{g'(1-g-g')} & \text{if } g \le \dfrac{(1-g')^2}{2-g'} \\[3mm] \dfrac{1-g'}{g'} & \text{if } g \ge \dfrac{(1-g')^2}{2-g'}. \end{cases}$$

The above formulas allow us to define a function $f \colon (g, g') \to (\tau, \tau')$ by the formula

$$f(g,g') = \begin{cases} \left(\dfrac{g'}{g(1-g-g')}, \dfrac{g}{g'(1-g-g')} \right) & \text{if } g' \le \dfrac{(1-g)^2}{2-g} \text{ and } g \le \dfrac{(1-g')^2}{2-g'} \\[4mm] \left(\dfrac{g'}{g(1-g-g')}, \dfrac{1-g'}{g'} \right) & \text{if } g' \le \dfrac{(1-g)^2}{2-g} \text{ and } g \ge \dfrac{(1-g')^2}{2-g'} \\[4mm] \left(\dfrac{1-g}{g}, \dfrac{g}{g'(1-g-g')} \right) & \text{if } g' \ge \dfrac{(1-g)^2}{2-g} \text{ and } g \le \dfrac{(1-g')^2}{2-g'} \\[4mm] \left(\dfrac{1-g}{g}, \dfrac{1-g'}{g'} \right) & \text{if } g' \ge \dfrac{(1-g)^2}{2-g} \text{ and } g \ge \dfrac{(1-g')^2}{2-g'} \end{cases}$$

Finding the image of f will tell us what are all the possible pairs of thicknesses (τ, τ') that can be realized by self-similar pairs of Cantor sets with only one overlapped point in their intersection. The domain of f and the four domains of definition it is divided into are shown in Figure 15. Notice that the two curves that divide the domain of f into four subdomains intersect at the point $(1 - 1/\sqrt{2}, 1 - 1/\sqrt{2})$. Also notice that the image of this point under f is the point $(\tau, \tau') = (1 + \sqrt{2}, 1 + \sqrt{2})$.

We will find the image of f by first computing the curve in the domain of f upon which τ is constant (i.e., a level set of the first component of f) and then maximizing τ' on this curve of constant τ.

A curve of constant τ is defined by the equation

$$\frac{g'}{g(1-g-g')} = \tau \qquad \text{when} \qquad g' \le \frac{(1-g)^2}{2-g}$$

which simplifies to

$$g' = \frac{g\tau(1-g)}{1+g\tau}$$

and also by the equation

$$\frac{1-g}{g} = \tau \qquad \text{when} \qquad g' \ge \frac{(1-g)^2}{2-g}$$

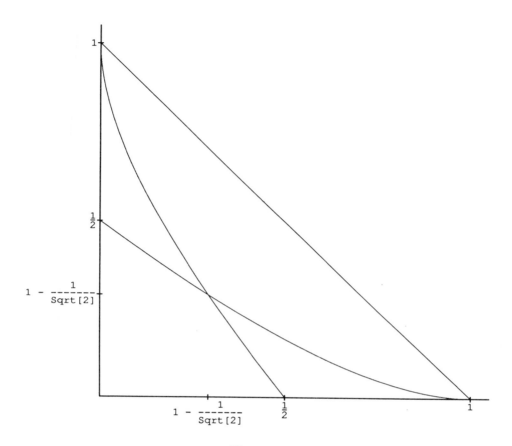

Figure 15

which simplifies to

$$g = \frac{1}{1+\tau}.$$

Pictures of this level set for values of τ equal to $1/2$, 1, 2 and 5 are given in Figure 16.

Similarly, curves of constant τ' (i.e., the level curves of the second component of f) are given by

$$g = \frac{g'\tau'(1-g')}{1+g'\tau'} \qquad \text{when} \qquad g \le \frac{(1-g')^2}{2-g'}$$

and

$$g' = \frac{1}{1+\tau'} \qquad \text{when} \qquad g \ge \frac{(1-g')^2}{2-g'}.$$

Figure 17 shows the curves of constant τ and τ' drawn together (with $\tau = 5$ and $\tau' = 2$).

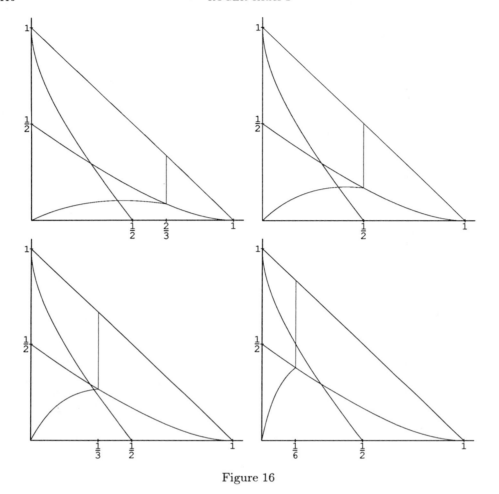

Figure 16

It is not very difficult to show that the maximum value of τ' on the curve of constant τ must be attained on one of the curves that defines the four regions of the domain of f, i.e., on one of the two curves

$$g' = \frac{(1-g)^2}{2-g} \qquad \text{or} \qquad g = \frac{(1-g')^2}{2-g'}.$$

This fact can be seen geometrically by looking at the shapes of the level sets of the first and second coordinates of f (i.e., the curves of constant τ and τ') or by restricting the function f to the curve of constant τ and then maximizing the resulting function of one variable for τ'.

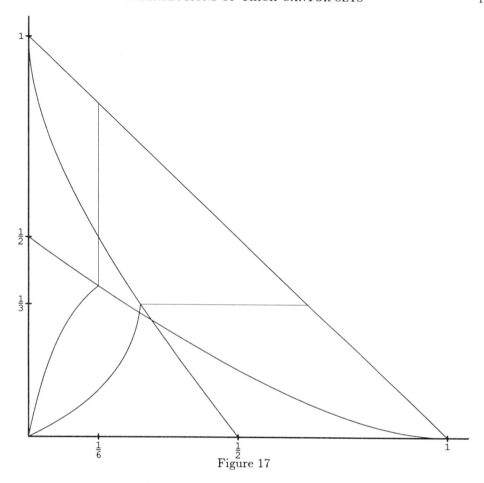

Figure 17

So to find the image of f, we need to find the image under f of the two curves

$$g' = \frac{(1-g)^2}{2-g} \qquad \text{and} \qquad g = \frac{(1-g')^2}{2-g'}.$$

First let's look at the image under f of the curve

$$g' = \frac{(1-g)^2}{2-g} \quad \text{with } g \geq 1 - \frac{1}{\sqrt{2}}. \tag{1}$$

On this curve we have

$$\tau = \frac{1-g}{g} \qquad \text{and} \qquad \tau' = \frac{1-g'}{g}$$

which is equivalent to

$$g = \frac{1}{1+\tau} \qquad \text{and} \qquad g' = \frac{1}{1+\tau'}.$$

Plugging these values for g and g' into the curve (1) gives

$$\frac{1}{1+\tau'} = \frac{\left(1 - \dfrac{1}{1+\tau}\right)^2}{2 - \dfrac{1}{1+\tau}} = \frac{\tau^2}{(1+\tau)(1+2\tau)}$$

which simplifies to

$$\tau' = \frac{\tau^2 + 3\tau + 1}{\tau^2}$$

with $\tau \leq 1 + \sqrt{2}$. Similarly, the image under f of the curve

$$g = \frac{(1 - g')^2}{2 - g'} \quad \text{with } g' \geq 1 - \frac{1}{\sqrt{2}} \tag{2}$$

is the graph of the equation

$$\tau = \frac{\tau'^2 + 3\tau' + 1}{\tau'^2} \quad \text{with } \tau' \leq 1 + \sqrt{2}.$$

Now we shall compute the image under f of the curve

$$g' = \frac{(1 - g)^2}{2 - g} \quad \text{with } g \leq 1 - \frac{1}{\sqrt{2}}. \tag{3}$$

On this curve we have

$$\tau = \frac{1 - g}{g} \quad \text{and} \quad \tau' = \frac{g}{g'(1 - g - g')}.$$

So we know that

$$g = \frac{1}{1+\tau} \quad \text{and} \quad g' = \frac{(1 - g)^2}{2 - g}$$

which is the same as

$$g = \frac{1}{1+\tau} \quad \text{and} \quad g' = \frac{\tau^2}{(1+\tau)(1+2\tau)}.$$

Plugging these values for g and g' into the second coordinate of f we get

$$\tau' = \frac{\dfrac{1}{1+\tau}}{\dfrac{\tau^2}{(1+\tau)(1+2\tau)}\left(1 - \dfrac{1}{1+\tau} - \dfrac{\tau^2}{(1+\tau)(1+2\tau)}\right)}$$

which simplifies to

$$\tau' = \frac{(1+2\tau)^2}{\tau^3}$$

with $\tau \geq 1 + \sqrt{2}$. Similarly, the image under f of the curve

$$g = \frac{(1 - g')^2}{2 - g'} \quad \text{with } g' \leq 1 - \frac{1}{\sqrt{2}} \tag{4}$$

is the graph of the equation

$$\tau = \frac{(1 + 2\tau')^2}{\tau'^3} \quad \text{with } \tau' \geq 1 + \sqrt{2}.$$

References

[B]   R. P. Boas, Jr., *A Primer of Real Functions*, Mathematical Association of America, Washington, D.C., 1966.

[GH]   J. Guckenheimer, and P. Holmes, *Nonlinear Oscillations, Dynamical Systems, and Bifurcation of Vector Fields*, Springer-Verlag, New York, 1983.

[HKY] B.R. Hunt, I. Kan and J. A. Yorke, *When Cantor sets intersect thickly*, Trans. Amer. Math Soc. (to appear).

[N]   S. E. Newhouse, *Lectures on dynamical systems*, Dynamical Systems, C. I. M. E. Lectures, Bressanone, Italy, June, 1978, Progress in Mathematics, No. 8, Birkhäuser, Boston, 1980, pp. 1–114.

[PT]   J. Palis, and F. Takens, *Homoclinic Bifurcations and Hyperbolic Dynamics*, Chapter 4, Cantor Sets in Dynamics–Fractional Dimensions (to appear).

[Ro]   C. Robinson, *Bifurcation to infinitely many sinks*, Commun. Math. Phys. **90** (1983), 433–459.

[R]   D. Ruelle, *Elements of Differentiable Dynamics and Bifurcation Theory*, Academic Press, New York, 1989.

[W]   R. F. Williams, *How big is the intersection of two thick Cantor sets?* Continuum Theory and Dynamical Systems (M. Brown, ed.), Proc. Joint Summer Research Conference on Continua and Dynamics (Arcata, California, 1989), Amer. Math. Soc., Providence, R.I., 1991.

[W2]   _____, *Geometric theory of dynamical systems*, Workshop On Dynamical Systems (Z. Coelho, ed.), Proc. of the conference at the International Centre for Theoretical Physics (Trieste, 1988), Pitman Research Notes in Mathematics Series 221, Longman, New York, 1990, pp. 67–75.

Department of Mathematics and Statistics, Case Western Reserve University, Cleveland, Ohio

Editorial Information

To be published in the *Memoirs*, a paper must be correct, new, nontrivial, and significant. Further, it must be well written and of interest to a substantial number of mathematicians. Piecemeal results, such as an inconclusive step toward an unproved major theorem or a minor variation on a known result, are in general not acceptable for publication. *Transactions* Editors shall solicit and encourage publication of worthy papers. Papers appearing in *Memoirs* are generally longer than those appearing in *Transactions* with which it shares an editorial committee.

As of March 1, 1992, the backlog for this journal was approximately 9 volumes. This estimate is the result of dividing the number of manuscripts for this journal in the Providence office that have not yet gone to the printer on the above date by the average number of monographs per volume over the previous twelve months. (There are 6 volumes per year, each containing about 3 or 4 numbers.)

A Copyright Transfer Agreement is required before a paper will be published in this journal. By submitting a paper to this journal, authors certify that the manuscript has not been submitted to nor is it under consideration for publication by another journal, conference proceedings, or similar publication.

Information for Authors

Memoirs are printed by photo-offset from camera copy fully prepared by the author. This means that the finished book will look exactly like the copy submitted.

The paper must contain a *descriptive title* and an *abstract* that summarizes the article in language suitable for workers in the general field (algebra, analysis, etc.). The *descriptive title* should be short, but informative; useless or vague phrases such as "some remarks about" or "concerning" should be avoided. The *abstract* should be at least one complete sentence, and at most 300 words. Included with the footnotes to the paper, there should be the 1991 *Mathematics Subject Classification* representing the primary and secondary subjects of the article. This may be followed by a list of *key words and phrases* describing the subject matter of the article and taken from it. A list of the numbers may be found in the annual index of *Mathematical Reviews*, published with the December issue starting in 1990, as well as from the electronic service e-MATH **[telnet e-MATH.ams.com (or telnet 130.44.1.100)**. Login and password are **e-math]**. For journal abbreviations used in bibliographies, see the list of serials in the latest *Mathematical Reviews* annual index. Authors are encouraged to supply electronic addresses when available. These will be printed after the postal address at the end of each article.

Electronically-prepared manuscripts. The AMS encourages submission of electronically-prepared manuscripts in $\mathcal{A}_{\mathcal{M}}\mathcal{S}$-TEX or $\mathcal{A}_{\mathcal{M}}\mathcal{S}$-LATEX. To this end, the Society has prepared "preprint" style files, specifically the amsppt style of $\mathcal{A}_{\mathcal{M}}\mathcal{S}$-TEX and the amsart style of $\mathcal{A}_{\mathcal{M}}\mathcal{S}$-LATEX, which will simplify the work of authors and of the production staff. Those authors who make use of these style files from the beginning of the writing process will further reduce their own effort.

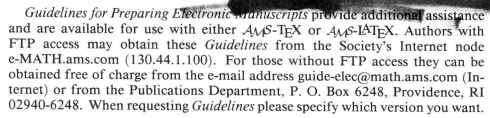

Guidelines for Preparing Electronic Manuscripts provide additional assistance and are available for use with either $\mathcal{A}_{\mathcal{M}}\mathcal{S}$-TEX or $\mathcal{A}_{\mathcal{M}}\mathcal{S}$-LATEX. Authors with FTP access may obtain these *Guidelines* from the Society's Internet node e-MATH.ams.com (130.44.1.100). For those without FTP access they can be obtained free of charge from the e-mail address guide-elec@math.ams.com (Internet) or from the Publications Department, P. O. Box 6248, Providence, RI 02940-6248. When requesting *Guidelines* please specify which version you want.

Electronic manuscripts should be sent to the Providence office only after the paper has been accepted for publication. Please send electronically prepared manuscript files via e-mail to pub-submit@math.ams.com (Internet) or on diskettes to the Publications Department address listed above. When submitting electronic manuscripts please be sure to include a message indicating in which publication the paper has been accepted.

For papers not prepared electronically, model paper may be obtained free of charge from the Editorial Department at the address below.

Two copies of the paper should be sent directly to the appropriate Editor and the author should keep one copy. At that time authors should indicate if the paper has been prepared using $\mathcal{A}_{\mathcal{M}}\mathcal{S}$-TEX or $\mathcal{A}_{\mathcal{M}}\mathcal{S}$-LATEX. The *Guide for Authors of Memoirs* gives detailed information on preparing papers for *Memoirs* and may be obtained free of charge from AMS, Editorial Department, P. O. Box 6248, Providence, RI 02940-6248. The *Manual for Authors of Mathematical Papers* should be consulted for symbols and style conventions. The *Manual* may be obtained free of charge from the e-mail address cust-serv@math.ams.com or from the Customer Services Department, at the address above.

Any inquiries concerning a paper that has been accepted for publication should be sent directly to the Editorial Department, American Mathematical Society, P. O. Box 6248, Providence, RI 02940-6248.

This journal is designed particularly for long research papers (and groups of cognate papers) in pure and applied mathematics. Papers intended for publication in the *Memoirs* should be addressed to one of the following editors:

Ordinary differential equations, partial differential equations, and applied mathematics to JOHN MALLET-PARET, Division of Applied Mathematics, Brown University, Providence, RI 02901-9000

Harmonic analysis, representation theory and Lie theory to AVNER D. ASH, Department of Mathematics, The Ohio State University, 231 West 18th Avenue, Columbus, OH 43210

Abstract analysis to MASAMICHI TAKESAKI, Department of Mathematics, University of California at Los Angeles, Los Angeles, CA 90024

Real and harmonic analysis to DAVID JERISON, Department of Mathematics, M.I.T., Rm 2–180, Cambridge, MA 02139

Algebra and algebraic geometry to JUDITH D. SALLY, Department of Mathematics, Northwestern University, Evanston, IL 60208

Geometric topology, hyperbolic geometry, infinite group theory, and general topology to PETER SHALEN, Department of Mathematics, University of Illinois at Chicago, Chicago, IL 60680

Algebraic topology and differential topology to RALPH COHEN, Department of Mathematics, Stanford University, Stanford, CA 94305

Global analysis and differential geometry to ROBERT L. BRYANT, Department of Mathematics, Duke University, Durham, NC 27706-7706

Probability and statistics to RICHARD DURRETT, Department of Mathematics, Cornell University, Ithaca, NY 14853-7901

Combinatorics and Lie theory to PHILIP J. HANLON, Department of Mathematics, University of Michigan, Ann Arbor, MI 48109-1003

Logic, set theory, general topology and universal algebra to JAMES E. BAUMGARTNER, Department of Mathematics, Dartmouth College, Hanover, NH 03755

Algebraic number theory, analytic number theory, and automorphic forms to WEN-CHING WINNIE LI, Department of Mathematics, Pennsylvania State University, University Park, PA 16802-6401

Complex analysis and nonlinear partial differential equations to SUN-YUNG A. CHANG, Department of Mathematics, University of California at Los Angeles, Los Angeles, CA 90024

All other communications to the editors should be addressed to the Managing Editor, JAMES E. BAUMGARTNER, Department of Mathematics, Dartmouth College, Hanover, NH 03755.